The Rainforests of Cameroon

The Rainforests of Cameroon

Experience and Evidence
from a Decade of Reform

Giuseppe Topa
Alain Karsenty
Carole Megevand
Laurent Debroux

THE WORLD BANK
Washington, DC

PROFOR

ISBN: 978-0-8213-7878-6
eISBN: 978-0-8213-7937-0
DOI: 10.1596/ 978-0-8213-7878-6

Cover photo: © George Steinmetz/Corbis

Cataloging-in-Publication data for this title is available from the Library of Congress.

Contents

Boxes

Figures

Tables

Foreword

It is a great pleasure to introduce this assessment of Cameroon's forest sector reforms. The book testifies to our achievements, weighs emerging challenges, and highlights the way forward. It comes at an important juncture, when reforms must gain new momentum.

Our forests are a national and global treasure to be safeguarded and sustained. The reforms described were designed based on the fundamental principle that forests, when properly managed, will sustain our economy, improve the welfare of our poorest citizens, and protect the environment and biodiversity into the future. Our convictions have driven a challenging and often unpopular reform agenda that strikes at the heart of the vested interests that have long defined the sector—industrial, political, administrative. As a result, we are reinstating rule of law and placing more decision-making power in the hands of civil society and local government institutions. Thus, our work concerns more than the nation's forests. It promotes good governance, democratization, and decentralization.

Reform on this scale is not implemented overnight. That is why the long-term perspective is essential to understanding the reform process. Setbacks, impediments, and resistance are to be expected—and we have encountered them. Even now, our forests and wildlife remain the targets of illegal activity. Opportunists can still deprive our nation and our forest communities of income and resources.

Yet reform has progressed steadily. Support from the donor community, including the World Bank, has strengthened the hand of reformers in government and in civil society. These partnerships have been—and continue to be—integral to progress. We have started in earnest to conserve biodiversity, implement forest management plans, and create community/local council forests. A transparent system for awarding logging concessions is in place, providing incentives to manage concessions responsibly and to curb illegal logging. Under a new taxation system, Cameroon obtains more revenue from forest resources; some of those financial resources are returned to local communities for development. We are improving the traceability of forest products and limiting opportunities for corruption.

This book may not always be flattering to the Government of Cameroon. I do not necessarily endorse all of the statements and conclusions. But I recognize that progress requires serious analysis and information, multiple voices, and some disagreement. I am proud to introduce this document because, all in all, it gives a fair account of our struggles and achievements, and it suggests solutions that build on past accomplishments. It shows that the keys to progress exist, are in our hands, and will be used to succeed.

Ephraïm Inoni
Prime Minister of Cameroon

Preface and Acknowledgments

In 1994, the Government of Cameroon introduced an array of forest policy reforms, both regulatory and market-based, to support a more organized, transparent, and sustainable system for accessing and using forest resources. This report describes how these reforms played out in the rainforests of Cameroon. The intention is to provide a brief account of a complex process and identify what worked, what did not, and what can be improved.

Although the reforms originated and evolved in this specific context, lessons from this experience may prove useful to countries facing similar circumstances, especially forest-rich countries with weak governance. In particular, it is hoped that the results of this inquiry prove useful to the Government, people, and development partners of Cameroon; policy experts; development practitioners; students; and persons with a more general interest in forests and in forest and environmental policy.

This report and more generally the reform process it describes have benefited greatly from the vision, insight, and hard work of people in the Cameroonian Government and civil society as well as managers and specialists in the World Bank and collaborating institutions.

It is unlikely that the reforms described in this report would have progressed without the efforts of the Government Committee in charge of structural reforms. During its tenure from 2000 to 2004, the Committee considerably broadened support for the reforms. Thanks are also due to several unnamed members of the Presidency, who ensured communication at the highest political level on key reform issues and increased the momentum for change, and to many unnamed members of other institutions.

Special thanks go to Prime Minister Ephraïm Inoni, who organized the first Heads of State Forum on Forests in 1999 and was among first to understand the strategic importance of forests in building trust between the state and its citizens and shaping Cameroon's international image. Additionally, Denis Koulagna (former Director, Wildlife, and currently General Secretary of the Ministry of Forests) and Hyrceinthe Bengono (former Director, Forests, and currently Head of Community Forestry) had the vision and courage to foster the implementation of new, and at times unpopular, measures.

It is also essential to acknowledge Cameroon's chapter of the Network of Parliamentarians for Sustainable Forest Ecosystem Management. These MPs worked hard to keep forest issues in the public eye and provided reviews of the draft report.

At the World Bank, thanks are due first of all to Mohammed Bekhechi (Lead Legal Counsel), Brendan Horton (Lead Economist), Florence Charlier (Senior Economist), David Tchuinou (Senior Economist), Clotilde Ngomba (Senior Agricultural Economist), and Serge Menang (Forest Specialist). Intervening at different stages of the reform process, these experts contributed invaluable legal and economic expertise, local knowledge, and infinite passion. Joseph Baah-Dwomoh (former Sector Manager), Menachem Katz and Mbuyamu Matungulu (former IMF Chief of Mission and Resident Representative for Cameroon), Ali Khadr (former Country Director), Bob Lacey and Madani Tall (former Country Managers for Cameroon), and Abdoulaye Seck (Senior Economist) made important contributions as well.

The multilateral and bilateral partners of Cameroon in the forest sector, especially the UK Department for International Development (DFID), the Canadian International Development Agency (CIDA), the Agence Françoise de Developpement (AFD) and the French Ministry of Cooperation, the Deutsche Gesellschaft für Technische Zusammenarbeit (GTZ), the Netherlands, and the European Commission, individually and collectively helped the Government of Cameroon conceive and implement parts of the reforms.

Colleagues at WWF, IUCN, WCS, CED, SNV, and many other national and international NGOs played a vital role in espousing and sustaining the spirit of reform in collaboration with their constituents in Cameroon's civil society.

We would like to thank the multi-donor Program on Forests (PROFOR) for their financial and communications support towards the publication, translation, and dissemination of this report.

We also wish to thank the many colleagues who reviewed drafts of this report, especially those who helped us strengthen the report by providing valuable criticism. Special thanks are due to Uma Lele, Frances Seymour and Paolo Cerruti (CIFOR), Reiner Tegtmeier (GW), Stuart Wilson, Albert Barume, and Guy Huot (REM), Pascal Cuny (SNV), Ofir Drori (Last Great Ape Association), Pierre Methot (WRI), Roger Fotso (WCS-Cameroon), Hans Schipulle (GTZ), and Mikael Grut.

The authors of the book are Giuseppe Topa, longtime Lead Forest Specialist for Africa with the World Bank; Alain Karsenty, Research Economist with the Centre de coopération internationale en recherche agronomique pour le développement (CIRAD) Montpellier; and Carole Megevand (Natural Resources Specialist) and Laurent Debroux (Senior Natural Resources Specialist), who both spent the first part of their careers with the World Bank working on the forest reform program in Cameroon.

Abbreviations

AAC	annual allowable cut (assiette annuelle de coupe)
ANAFOR	National Forest Development Support Agency (Agence Nationale d'Appui au Développement Forestier)
ATO	African Timber Organization
BBNRMC	Bimbia Bonadikombo Natural Resource Management Council
CARPE	Central African Program for the Environment
CCU	Central Control Unit (CCU)
CED	Center for Environment and Development (Centre pour l'Environnement et le Développement)
CEFDHAC	Conference on Central African Moist Forest Ecosystems
CENADEFOR	National Center for Forestry Development (Centre National de Développement des Forêts
CFAF	CFA franc
CIRAD	French Agricultural Research Centre for International Development (Centre de Coopération Internationale en Recherche Agronomique pour le Développement)
CITES	Convention on International Trade in Endangered Species

COMIFAC	Conference of Ministers in Change of Forests in Central Africa
CTS	Technical Oversight Committee for Economic Reforms (Comité Technique de Suivi des Réformes Economiques)
DFID	Department for International Development (UK)
DRC	Democratic Republic of Congo
€	Euro
EAP	Emergency Action Plan
ECOFAC	Ecosystèmes Forestiers d'Afrique Centrale
ESI	Environmental Sustainability Index
EU	European Union
FAO	Food and Agriculture Organization of the United Nations
FESP	Forest and Environment Sector Program
FLEGT	Forest Law Enforcement, Governance, and Trade
FMU	Forest Management Unit
FOB	free on board
FOCARFE	Cameroonian Foundation for Concerted Action and Training on the Environment (Fondation Camerounaise d'Actions Rationalisées et de Formation sur l'Environnement)
FSC	Forest Stewardship Council
GDP	Gross domestic product
GFW	Global Forest Watch
GLIN	Global Legal Information Network
GTZ	German Technical Cooperation (Deutsche Gesellschaft für Technische Zusammenarbeit)
GW	Global Witness
ha	hectare
HIPC	Heavily Indebted Poor Country
I&D	Institutions et Développement
IDA	International Development Association
IMF	International Monetary Fund
IPDP	Indigenous Peoples Development Plan
ISO	International Organization for Standardization
ITTO	International Tropical Timber Organization
IUCN	International Union for the Conservation of Nature
LAGA	Last Great Ape Organization
m^3	cubic meter

MDC	minimum diameter for cutting
MINEF	Ministry of Environment and Forests (Ministère de l'Environnement et des Forêts)
MINFOF	Ministry of Forests and Wildlife (Ministère des Forêts et de la Faune)
MINEP	Ministry of the Environment and Protection of Nature (Ministère de l'Environnement et de la Protection de la Nature)
NEPAD	New Partnership for Africa's Development
NGO	nongovernmental organization
OCFSA	Organization for the Conservation of African Wildlife (Organisation pour la Conservation de la Faune Sauvage d'Afrique)
ONADEF	National Office of Forest Development (Office National de Développement des Forêts)
ONAREF	National Bureau for Forest Regeneration (Office National de Régénération des Forêts)
PAME	Protected Area Management Effectiveness
PCI	Principles, Criteria, and Indicators
PCGBC	Cameroon Biodiversity Conservation and Management Program (Programme de Conservation et de Gestion de la Biodiversité au Cameroun)
PRGF	Poverty Reduction and Growth Facility
PSRF	Enhanced Forest Revenue Program (Programme de Sécurisation des Recettes Forestières)
REDD	Reduction of Emissions from Deforestation and Degradation
REM	Resource Extraction Monitoring
RIL	Reduced Impact Logging
SAC	Structural Adjustment Credit
SIGIF	Système Informatique de Gestion de l'Information Forestière
SNV	Netherlands Development Organisation
SOFIBEL	Bélabo lumber products corporation (Société Forestière et Industrielle de Bélabo)
TFAP	Tropical Forest Action Plan
UFA	forest management unit (unités forestières d'aménagement)
UK	United Kingdom
UNDP	United Nations Development Programme

US$	U.S. dollar
VC	ventes de coupe
WRI	World Resources Institute
WWF	World Wildlife Fund

All dollar amounts are given in U.S. dollars unless otherwise specified.

Exchange rate for the CFA franc (CFAF):
Before 1994: $1 = CFAF 250
After 1994: $1 = CFAF 500

Overview

Cameroon's forest sector has undergone reforms of unprecedented scope and depth.

The barriers to placing Cameroon's forests at the service of its people, its economy, and the environment originated with the extractive policies of successive colonial administrations. The barriers were further consolidated after independence through a system of political patronage and influence in which forest resources became a coveted currency for political support.

These deeply entangled commercial and political interests have only recently—and reluctantly—started to diverge. In 1994, the government introduced an array of forest policy reforms, both regulatory and market based. The reforms changed the rules determining who could gain access to forest resources, how access could be obtained, how those resources could be used, and who would benefit from their use.

This report assesses the outcomes of reforms in forest-rich areas of Cameroon, where the influence of industrial and political elites has dominated since colonial times.

The reforms sought to replace chaotic and often secretive arrangements for accessing forest resources with a more organized, transparent, and sustainable system that would benefit greater numbers of people and the environment. In essence, the story of Cameroon's forest reforms is largely one of regulating access to natural resources; balancing public and private interests; and integrating wider economic, cultural, and environmental perspectives of the value of forests.

The reforms' initial focus on the most powerful actors in the forest sector made it possible to address other long-neglected issues more effectively.

The forest industry's ubiquitous presence in the rainforests and its ties to economic and political elites made it the obvious entry point for reform. The reforms had to deal with the expanding privilege of industry and the role of forest institutions or risk becoming irrelevant. The success of broader and more effective interventions pertaining to forest communities and cultures, forest products, and environmental issues rested on the ability to rein in the timber industry and to establish clear parameters and standards of behavior for all stakeholders.

Transparency, public information, and public participation built awareness and support for the reforms.

New technologies such as geographic information systems and the Internet fueled an increasingly public discourse on Cameroon's forests, accelerating the pace of reform. Improved systems to track logging and changes in forest cover—some of them used in control operations by internationally respected nongovernmental organizations that act as independent observers—heightened pressure on public officials to enforce forest regulations. Public disclosure of forest violations in national newspapers and international bulletins brought increased transparency and accountability to a sector plagued by corruption (a conclusion reinforced in a recent Transparency International report).

The new forest regulations called for participation and public consultation, which gradually transformed latent demand for good governance into expressed demand. The people affected by forest policies—and a growing number of civil society organizations—can now obtain, use, and contribute to an increasing wealth of public information to clarify their rights and advance their agendas.

Starting in southern Cameroon, forest zoning set the stage for more effective land management.

Zoning and gazetting introduced regulation and clarity into a chaotic system in which the government had acted as the landlord of the forest.

**Cameroon's forest landscape permeates cultural identities, local liveli-
hoods, national economic interests, and global ecological stability.** The
forest's significance for biodiversity and the environment is enormous. At the
household level, the forest directly provides about 8 million rural and poor Cameroonians with traditional medicines, important complements to the staple
diet, domestic energy, and construction material.

Providing up to 13,000 formal and perhaps 150,000 informal jobs, the forest
sector is Cameroon's largest employer outside the public sector and its second-
largest source of export revenue after petroleum, accounting for 29 percent of
nonpetroleum exports in 2001 and for 26 percent in 2004. Forestry and related
activities (including informal activities) accounted for 4.8 percent of nonpetroleum gross domestic product in 2004.

Not surprisingly, the value of timber has dominated perceptions of the forest's value, but this perception is not accurate. The total value of forest products for which trade statistics or estimates exist—timber, charcoal, okok leaves
(*Gnetum* spp.), gum arabic, and *Prunus africana* bark—is about $580 million. Of
this, $120 million is derived from products other than timber. Yet many nontim-
ber forest products are not formally traded, and their overall value has not been
quantified. The value of domestic energy alone is $130 million, however, and if
bush meat, fruit, thatching, and medicinal plants are considered along with the
value of biodiversity conservation and environmental services, the total value of
nontimber forest products will well exceed that of traditional products. Nontimber forest products, like other forest products, are subject to illegal exploitation,
and the consequences can be just as socially and environmentally devastating.

Zoning forests into protected, commercial, and community or local council zones provided a foundation for protecting forest land, determining
how it would be managed, and specifying which forest areas could be converted to other uses. Together, zoning and gazetting enabled stakeholders (government, communities, industry, and so on) to establish secure
use rights, which had been lacking since colonial times. The zoning plan
now covers some 14 million hectares and is being expanded throughout
Cameroon. The plan is proving effective at staving off conflict over trees
and land. Even so, it is clear now that there were significant omissions,
such as insufficient recognition of the rights of Pygmy communities and
others living in the forest—an important lesson for Cameroon and other
countries in the region.

Forest harvest rights are now granted in a more regulated and public way.

The types of industrial harvesting rights and the way they are granted are fundamentally different now as a result of the reforms. Long-term rights to forest management units (*unités forestières d'aménagement*, or UFAs), which are required to be harvested selectively and managed sustainably, have replaced short-term permits to harvest both small and large forests.

These harvesting rights are no longer awarded at the discretion of government officials. A new public auction system, overseen by an independent observer for the allocation of titles, requires the winning firm to prove it possesses sound technical capacity to manage a UFA and to offer the highest bid. Firms with a record of forest infractions are excluded from bidding. Short-term permits to harvest and sell timber from much smaller areas than UFAs, known as *ventes de coupe*, are also issued by auction.

Competition engendered by the auctions has helped increase the sums that companies pay to access forests (the area tax) from a base of $0.60 per hectare per year in 1990 (set administratively) to averages of $5.60 per hectare per year for UFAs in 2006 and $13.70 per hectare for ventes de coupe in 2005. The bidding process attracted new and more environmentally and fiscally responsible investors to Cameroon.

Community and local governing councils' rights to manage forests have also been established within limits defined by the zoning process.

The reforms have brought significant achievements.

- *A relatively well-conserved forest resource.* With more than 60 percent of its forests under biodiversity conservation, forest management plans, and community and local council forests, Cameroon is well set to maintain a permanent forest estate of 8 million hectares. Illegal logging in permanent production forests has been reduced, although it persists in rural areas. The combined timber harvest from legal sources is estimated to have fallen by about 1 million cubic meters compared with 1997, following constraints imposed by the implementation of the new management plans. Degradation of the permanent forest is declining. Illegal logging by small operators has emerged as a new and problematic trend. Even so, the main drivers of deforestation are the expansion of subsistence agriculture, insecure land tenure, and other issues outside the forest sector—where the primary solutions also reside.

- *Deforestation has been contained.* The most recent and authoritative studies on deforestation conducted by the European Commission's

Joint Research Centre, the Université Catholique de Louvain, and the University of South Dakota (de Wasseige and others 2009) show that average gross deforestation between 1990 and 2005 remained modest, around 0.14 percent per year. And the use of more refined sensing technology allowed reassessment to 0.14 percent per year, down from 0.28 percent, of the rate for gross deforestation previously reported for the 1990–2000 period.[1]

- *Improved and internationally recognized management practices in a restructured forest industry.* A significant part of industry has restructured, adapting its business model to cost structures that recognize higher value in the resource base, new investments for management plans, and increased social and environmental responsibility. By 2006, 55 UFAs, covering 4 million hectares (more than 71 percent of the UFA area on which harvesting rights had been awarded by 2005), had approved management plans. In that same year, concessions operating under management plans supplied more than 85 percent of commercial timber, up from less than 30 percent in 1998. Compliance with new regulations has helped reduce annual harvests and increasingly aligned harvests with the forest's capacity to regenerate. Although much remains to be done, the quality of management plans is considered better in Cameroon than in Southeast Asia and Brazil, where management norms are less precise and often not well enforced. Cameroon has the highest number of companies seeking certification from the Forest Stewardship Council.

- *Growing recognition of customary rights and the social welfare contributions of forests.* The reforms are explicitly linked with the recognition of customary rights of Bantu and Pygmy people. For example, for the first time, the traditional forest use rights of Baka, Bakola, and other people of the Pygmy group are acknowledged in forest regulations and contractual arrangements with forest concessionaires.

  Issues raised during environmental impact assessments have also started to be discussed publicly. Cameroon has established rules to preserve customary rights in all of its forests and is developing more detailed provisions to recognize and protect the rights of indigenous people.

  Greater regulation of industry, greater transparency in forest administration, and greater recognition of local claims to forests improved the prospects for community forests, which have progressed rapidly. In 1997 the first two community forests were established. By 2008, agreements had been concluded to create 135 forests on 621,246 hectares, and requests for an additional 700,000 hectares of community forests were being considered. Although neither the reformers nor the wider international community envisioned just how complex and conceptually challenging community forestry would prove to be, observers concur that community forestry in Cameroon is a decade ahead of efforts elsewhere in Africa's humid tropics, where the lessons from Cameroon are likely to be useful.

- *Institutionalized collaboration between forest institutions and civil society improved forest governance and transparency.* Collaboration with organizations acting as independent observers enabled the government and other partners to detect suspicious activity in concessions and parks and made inaction hard to justify. Independent observers reported on and were associated with the quality of auctions and forest control operations, and they helped to create a public register of forest and wildlife infractions.

An unfinished agenda, with high stakes, remains.

All in all, Cameroon has moved in the right direction. The reforms have created a framework for better forest governance and more sustainable management that is far more effective than the previous framework. They have helped to impose order over the most aggressively competing interests in the sector and have proven sufficiently adaptable to address deeper issues as well as emerging and changing concerns. Reforms must move forward, however, to encompass a significant unfinished agenda, with ultimately very high stakes.

- *Address the needs of indigenous peoples.* Special measures are required for indigenous peoples to participate in and benefit from forest reforms. The government's Indigenous Peoples Development Plan envisions a number of corrective actions for indigenous communities to attain more secure rights to land and resources, especially by conferring legal status on indigenous settlements, recognizing their traditional use rights, giving them greater access to markets for forest products, providing new community forests and hunting zones where those rights apply, ensuring greater representation of all marginalized groups within local forest management institutions, and improving their access to forest revenues for their benefit and that of the land they use. This effort will deepen the reforms' emphasis on recognizing the customary rights of all peoples who depend on Cameroon's forests, whatever their ethnic affiliation.

- *Give greater attention to nontimber forest products.* Nontimber forest products (such as medicinal plants, thatching, gum arabic, and bush meat) make a vital but still underappreciated contribution to the viability of forests and the national economy. These products are subject to uncontrolled and illegal exploitation, and arrangements to protect their sustained use must be considered in devising new arrangements to protect customary rights and encourage multiple, overlapping forest uses.

- *Attract new eco-investors to make conservation efforts sustainable.* Cameroon is moving to a more holistic approach in forest management, based on multiple uses of forests and integrated production of forest products and global environmental services. To this end, Cameroon seeks partners to transform about 1 million hectares of forest, originally destined for logging, into conservation concessions that will store carbon, harbor biodiversity, and yield other global goods. Arrangements for conservation concessions would need to include compensation for the economic and fiscal revenues forgone by removing the land from industrial use. Attracting qualified investors who can negotiate fair arrangements, especially with respect to enhancing local livelihoods, will be fundamental for Cameroon to retain a positive view of forest management options that are not based exclusively on timber.

- *Reshape community forestry.* The establishment of community and local council forests has expanded the role of communities in managing forests, yet often community forests have been diverted from their intended use and have not delivered the expected economic, social, and environmental benefits. Major adjustments need to be made, starting at the conceptual level, if community forests are to represent a better option than business-as-usual land and tree management. A first step is to remove well-known impediments, such as excessive regulation, and permit greater freedom in identifying the technologies and partnerships that can be used to manage community forests. Simplified procedures will be much more effective if communities improve their own capacity to implement them instead of deferring to other actors. A far broader step is to develop community initiatives that can be carried out throughout the forest space, whenever compatible with other forest uses (for example, in managed forest concessions and natural parks, where communities can partner with government, private operators, or nongovernmental organizations).

- *Give greater attention to impacts on local markets, small firms, and employment.* International markets and large firms may be financially important, but continued inattention to local markets and small firms has significant social implications and can undermine forest sector performance and governance in important ways. Local markets and small-scale enterprises are the source of numerous jobs—precarious but essential—for the poor, yet little effort has been made to strengthen this important forest constituency and enable it to operate openly. The challenges involved are not small. Distrust of bureaucratic processes and fear of harassment by officials discourage artisanal loggers from joining the formal sector. Given the chance to produce timber legally, they may simply sell it to the better-paying export industry rather than to local markets, which would create conditions for new illegal artisanal loggers to arise.

Fortunately, partnerships between small and large enterprises and innovative solutions to create a steady supply of material to undersupplied regions have started to emerge. They should be encouraged and replicated. Access to the more lucrative export market should continue to be subject to the law and to fair competition by all companies, large and small.

- *Adapt reform instruments to emerging forest management and industry needs.* While the new social, fiscal, and other obligations that accompany management plans are positive developments, integrating them into a new business model can be extremely demanding for industry. The fragility of Cameroon's industry as it moves toward a highly regulated and managed forest production system is underappreciated. High international timber prices have somewhat compensated for industry's reliance on a few tree species and sawn timber, but industry needs now to invest in more advanced processing and technologies to use a wider range of species and market a broader range of products.

The regulatory framework could be adapted to benefit both forest sustainability and industry. Rewarding responsible corporate behavior with more lenient bank guarantees and automatic reimbursement of value added tax may prove as important for conserving forests as punishing corporate misbehavior. The development of incentives for firms to gain international certification could improve the sustainability of forest resources and of individual firms' operations at the same time. The risks facing investors and concession holders need to be considered more seriously. Concession holders are required to absorb a large share of commodity price risk on the international timber market. Fine-tuning policies and adjusting taxation and incentives may be critical in helping firms cope with cyclical downturns in prices and demand.

Will the reforms and their lessons have lasting effects?

Ten years of reforms have laid a foundation that is not likely to vanish overnight—and that may ultimately find expression in settings other than Cameroon. Several elements of the reforms seem relatively durable, such as the recognition that, like the public treasury, national forests are public property and should be treated with comparable care and subject to comparable accountability systems and controls. Similarly, government and industry appear to have accepted that forests are a finite entity, that ensuring long-term access to this resource is costly, and that the environmental and financial sustainability of this resource depends on sustainable management and industry efficiency. They also appear to recognize that third-party verification and disclosure have considerable power to ensure accountability. Cameroon's civil society and parliament now expect to be involved in decisions pertaining to forests, just as communities expect to be involved in such decisions and to benefit from forests under community and industry management.

Government can go a long way toward sustaining the reforms by appropriating sufficient salaries and complementary funds to hire younger, better-trained Forest Administration personnel who can perform the more sophisticated regulatory and supervisory functions that are needed. With many Forest Administration staff retiring and significant resources becoming available from government and the donor community, little should hinder this important step.

Other elements of reform appear more vulnerable. Community forestry reforms are recent, weak, imperfect, and burdened by vested interests and unrealistic requirements. Opportunities for eco-investment to conserve the rainforests in a more stable and sustained manner require more attention from the donor community and other potential investors. The competitive allocation of forest harvesting rights could be threatened; all concessions will soon be allocated, and in the future the government may be uninterested in ensuring that auctions remain resistant to fraud. Continued engagement with reform-minded Cameroonians within and outside government is required to offset these risks.

Perhaps the most urgent question arising from Cameroon's experience with forest reform remains to be answered. Can reform in one sector change the wider trajectory of society in any meaningful way? Or will the forest sector remain at variance with the rest of the country until it succumbs to the inertia that prevails elsewhere? It is hoped that forest sector reform will rebound with renewed support—from government as well as civil society—and will strengthen broader efforts to promote good governance and stewardship throughout the country.

Note

1. 0.28 percent and 0.14 percent are gross deforestation rates; these rates are not affected by reforestation of deforested areas.

PART I

The Reforms and Their Setting

CHAPTER 1

The Historic and Environmental Context

The barriers to placing Cameroon's forests at the service of the country's people, its economy, and the environment have been under construction for more than 100 years. Commercial logging was initiated during the colonial period in the 1880s and expanded after the 1920s. Since independence in 1960, the interaction of commercial timber firms and government civil servants within Cameroon's system of political patronage and influence has shaped the role of forests in the country's economy. Since 1980, timber has been Cameroon's second-largest source of export revenue after petroleum, accounting for some 25 percent of the country's foreign exchange—far exceeding any other agricultural commodity.

These deeply entangled commercial and political interests have only recently—and reluctantly—started to diverge. The economic crisis of 1985 unleashed a series of changes, including structural adjustment, democratization, and decentralization, which inevitably affected forests and forest policy. More specifically, in 1994 the government introduced an array of forest policy reforms, both regulatory and market based. These reforms changed the rules determining who could gain access to forest resources, how those resources could be used, and who would ostensibly benefit from their use. As the reforms came into effect, long-standing economic, social, and environmental perspectives on forest resources began to shift in complex and not always predictable ways. More than a decade later, the reforms remain incomplete, but progress is evident in a sector regarded by many observers as virtually impervious to change, in a country that has been notorious for corruption.

The remainder of this chapter describes the context—Cameroon; its forests; and their economic, social, and political importance—in greater detail. The challenges, alliances, and conflicts that characterized the reform process, as well as the content of reforms, are discussed in the second and third chapters. The fourth chapter assesses the impacts of the reforms, and chapter 5 presents the data and analysis to support that assessment. Chapter 6 summarizes lessons, conclusions, and prospects for further reform.

The Natural and Cultural Setting

Cameroon's 16.5 million inhabitants occupy a territory of 475,000 square kilometers marked by profound changes in climate and terrain from the dry Sahelian north to the humid south. The country's ethnic and linguistic diversity—Cameroon claims more than 200 ethnolinguistic groups—is also considerable, and strong cultural identities are associated with the major ethnic groups and the ecological zones where they live. In the humid, forested south, the dominant Bantu ethnic groups are characterized by social structures that generally revolve around family clans with more diffuse authority. A number of marginalized indigenous groups also depend on the humid forests for their livelihoods, most notably the Pygmy peoples (Aka, Baka, Bakola, and others). Throughout Cameroon, political parties and associations have been superimposed on traditional structures. Recent attempts to decentralize administrative functions—however faltering—have imbued local authorities with a new prominence.

The Congo Basin rainforest, which covers more than 198 million hectares in six countries, including Cameroon, is "the second largest contiguous rainforest in the world after those of the Amazon Basin" (GFW 2000:9). Cameroon's share of these forests occupies 19.6 million hectares (CBFP 2006), about 40 percent of national territory. In surface area, this densely forested zone resembles that of Gabon (22 million hectares) and the Republic of Congo (22 million hectares), but its population density is much higher than in neighboring Equatorial Guinea, the Democratic Republic of Congo, Gabon, or the Republic of Congo, all of which are also in the moist equatorial zone.

Cameroon's rainforests (map 1; see map insert) have been described as "some of the Congo Basin's most biologically diverse and most threatened forests" (GFW 2000:8). They harbor a remarkable range of flora and fauna and provide about 8 million rural and poor Cameroonians with important nutritional complements, traditional medicines, domestic energy, and construction material. Their preservation is key to maintaining local cultures and global ecological stability.

The total value of forest products for which trade statistics or estimates exist—timber, charcoal, okok leaves (*Gnetum* spp.), gum arabic, and *Prunus africana* bark—is about $580 million, of which $120 million is derived from products other than timber (table 1.1). Many nontimber forest products are not formally traded, and their overall value has not been quantified. The value of domestic energy alone is estimated at $130 million, however, and if bush meat, fruits, thatching material, and medicinal plants are considered as well, the total value of nontimber forest products may well exceed that of traditional products.

Yet timber has dominated perceptions of the forest's value. Providing up to 13,000 formal and perhaps 150,000 informal jobs, the forest sector is Cameroon's largest employer outside the public sector and its second-largest source of export revenue after petroleum, accounting for 29 percent of nonpetroleum exports in 2001 and for 26 percent in 2004. Forestry and related activities (including activities in the informal sector) accounted for 4.8 percent of nonpetroleum gross domestic product in 2004.

Supplying building material to the domestic market is a largely informal and undocumented activity that removes an estimated 580,000–1,000,000 cubic meters of wood from Cameroon's forests each year.

Table 1.1. Value of some nontimber forest products

Product	Production	Value ($)
Gum arabic (*Acacia Senegal/seyal*)		
Official	300 t	600,000
Informal (mainly to Nigeria)	2,000 t	4,000,000
Okok (*Gnetum* spp.)	1,817 t (in 2005)	3,997,400
Prunus africana (African cherry) (bark)	1,750 t	1,750,000
Fuelwood	10 million t	70,000
Timber (domestic)	540,000 m^3	37,800,000
Timber (export)	2,300,000 m^3	460,000,000

Source: Official data on gum arabic from Mallet and others 2003; informal production estimate from A. Karsenty, personal communication; estimate for okok from GTZ and FAO (2007); estimate for *Prunus africana* from Chupezi and Ndoye 2006; estimate for fuelwood from A. Karsenty, personal communication (the total estimated value of fuelwood is $200 million, of which $130 million is collected and used within the household); and estimate for domestic timber from Cerutti and Tacconi 2006.

Although most of this timber is sold domestically, a significant proportion now makes its way to Nigeria and the Sahel through informal networks.

The pressure on forests for fuelwood is higher in Cameroon's northern forests than in the south, but it is likely to increase as urban areas and needs expand. The Food and Agriculture Organization of the United Nations (FAO) estimates that 9.5 million cubic meters of fuelwood are collected yearly in Cameroon (the estimate from the Ministry of Forests is even higher, at 12 million cubic meters). No data are available on its value or its contribution to household incomes, because gathering fuelwood is generally considered a domestic chore. No opportunity cost of labor is associated with it. Nor are data available on the opportunity costs associated with fuelwood collection, which can be considerable, particularly for women and children. On average, an urban household is thought to spend $50–55 each year on fuelwood; altogether, Cameroon's 1.3 million urban households probably spend about $65 million to $70 million on this resource annually.

Agriculture has also claimed its share of Cameroon's rainforests. From independence until the mid-1980s, the government encouraged the conversion of moist forests into small-scale coffee and cocoa farms. The incursion of these farms was facilitated by large commercial timber operations, which built roads and opened extensive tracts in remote areas. With the exception of Eastern Province, Cameroon's dense humid forest is now interspersed with small cocoa and coffee farms, in addition to the slash-and-burn plots of other crops maintained by local people.

Economic Depression and Recovery

Amid Cameroon's great diversity, poverty remains perhaps the greatest commonality. About 40 percent of the population lives under the poverty threshold of $1 per day, and Cameroon is off track for meeting most of the Millennium Development Goals (UNDP 2003). Cameroon was regarded as one of Africa's most vibrant economic successes until its economic and policy weaknesses were exposed in the mid-1980s, when sharply declining prices for coffee, cocoa, and oil led to a 60 percent decline in the external terms of trade. This severe shock, combined with an overvalued exchange rate, fiscal crisis, and economic mismanagement, was followed by years of prolonged economic stagnation and rapidly accumulating public debt.[1]

Although this debilitating depression came to an end after 1993, the economy has recovered only slowly, with moderate annual growth of 4.5 percent in real gross domestic product (GDP) and low inflation of

2 percent per year, based on a comprehensive reform agenda (which also changed the management and use of natural resources, especially forests). Cameroon became eligible for debt relief under the Heavily Indebted Poor Countries (HIPC) Initiative in October 2003; in April 2006 it completed the process, getting rid of part of its debt to the International Monetary Fund (IMF), the International Development Association (IDA), and the African Development Fund.

Since the economic recovery began in 1994, per capita GDP has reached only two-thirds of its predepression level, and most social indicators have not improved. Poverty and corruption remain widespread, although Cameroon's Transparency International corruption rating rose from last place in 1998 to 138th of 180 countries surveyed in 2007. This substantial achievement (discussed briefly in chapter 6) may reflect the effects of reforms described in this report (Transparency International 2007).

Environmental Stewardship and Protected Areas

The 1994 Forestry Law commits Cameroon to placing 30 percent of its surface area under protection—one of the highest proportions anywhere in the world. Cameroon's network of national parks, forest reserves, wildlife sanctuaries, zoological and botanical gardens, hunting zones, and community hunting zones represents about 17.6 percent of the nation's forest estate.[2] The government is still planning new protected areas in critical ecosystems: eleven new protected areas exceeding 700,000 hectares are under consideration, including a marine protected area near Kribi and a mangrove biodiversity reserve at Ndongoré.

On the 2005 Environmental Sustainability Index (ESI), which assessed the capacity of 146 nations to protect their natural resource endowments over the coming decades, Cameroon ranked 50th, with an index score of 52.5 out of 76.[3] Within the group of countries belonging to the New Partnership for Africa's Development (NEPAD), Cameroon ranked 7th. These relatively favorable scores reflect Cameroon's progress and commitment to protecting its biodiversity, although the challenges should not be underestimated, especially because Cameroon is still developing the capacity to manage its protected areas.

Forests and Cameroon's Political Economy

Cameroon became a German colony in 1884 and, following the First World War, was divided and ruled by Great Britain and France. French

Cameroon became an independent nation in 1960, merging with the British Cameroons in 1961. In 1972, under President Ahmadou Ahidjo, the country was unified and ruled through an autocratic, highly centralized, one-party system. The current president, Paul Biya, succeeded Ahidjo in 1982 and continued the one-party state until 1990, when a multiparty system was introduced. The first parliamentary elections took place in 1992, and presidential elections followed in 1997. Cameroon's government is nevertheless frequently characterized as authoritarian.

The forest sector's economic importance made it a valuable source of influence, political capital, and reward in Cameroon's extensive patronage system. Forest resources were used to reward political supporters as well as to mobilize support from rural communities for political appointments. The motives and behavior of various groups with an interest in Cameroon's forests are best understood in light of these circumstances. Vincent, Gibson, and Boscolo (2005) observed that forest resources, especially in the humid south, were the perfect currency for political support:

> Two reasons forests were attractive as a patronage good were that they were widely distributed throughout the country (at least in the southern part) and could be easily divided to reward multiple supporters (and differentiated by size to vary the rewards according to political importance). This made politicians from most of the country interested in how forest resources were allocated.

The Department of Forests, the gatekeeper to logging rights in Cameroon, focused on commercial timber and was widely seen as a lucrative posting for civil servants. Given its vital role in supporting the president, the Department of Forests enjoyed high-level access to his office but operated under quite direct presidential control. The initiation of democratic reforms has brought new complexity to forest management by involving a much broader range of actors with competing and overlapping interests and claims. Decisions about forest resources—such as the introduction of an auction system to bid for harvesting rights or the decision to ban log exports—involve significant changes in the relative costs and benefits for the stakeholders involved.

The forest industry is far more than a key stakeholder in Cameroon's forest sector. It is a power to be reckoned with in the national economy. On the ground it has a more extensive presence and is better organized than any government institution, and its long-standing connections with local administrative and traditional authorities have consistently sheltered it from modern and customary laws. Unlike prices of other agricultural exports, such as coffee and cocoa, timber prices have never collapsed but have, with occasional lows, risen steadily for 100 years.

European companies, principally French and Italian, traditionally have dominated Cameroon's forest industry. In remote rural areas, large foreign timber companies have acted as a surrogate state in providing services and infrastructure as well as in their informal political and economic dominance. They have interacted with domestic companies primarily by subcontracting logging rights and renting equipment. Over time these commercial relationships have deepened, in some instances to the point of shareholder participation. Starting in the 1980s, several Greek, Lebanese, and domestic companies entered the market. Although they were new to the industry and lacked processing capacity, these companies were attracted by the simplicity and profitability of operations and by their established links with political elites.

Institutions Involved in Forest Management

Cameroon's forest institutions were favorably positioned with respect to a valuable political and economic commodity, but they were not exempt from the economic trauma that spread throughout the country in the mid-1980s. Budgets declined precipitously in the Department of Forests, which became the Ministry of Environment and Forests (MINEF) in 1992 (and in 2004 split into the Ministry of Forests and Wildlife, MINFOF, and the Ministry of the Environment and the Protection of Nature, MINEP).

In 1992, 95 percent of MINEF's budget went to pay staff salaries, and only 5 percent was available for any other type of expenditure (WRI 2000). By 1994, the devaluation had reduced salaries of MINEF staff by 40 percent and further eviscerated their earning power. Employees were often paid months late and no longer received such attractive perquisites as vehicles and housing allowances. Ministry operating budgets had nearly vanished. The World Resources Institute (WRI 2000:64–5) summed up the bleak effects of these trends:

Declining salaries, poor working conditions, and the offer of very large sums of money provided a strong incentive for corruption. The average MINEF official earned CFAF 60,000 [$100] per month and had no means of transport or communication, but could gain millions of CFAF by not reporting logging in areas for which a company had no right.

As this passage indicates, it would have been extremely counterproductive to reinforce such institutions without reforming the rules by which they operated. Cameroon's forest institutions were weak, but this weakness was simply the manifestation of deeper political and institutional flaws in the forest sector.[4]

The core of the current forest development program is an effort to revitalize the human resources responsible for managing the green environment. No new staff members have been recruited to the nation's forest institutions since the late 1980s, leaving these institutions vulnerable to an array of new challenges in the forest sector. Under the National Forest and Environment Sector Program, forest institutions sought government support for a detailed strategy to recruit 1,550 new staff with specific qualifications between 2006 and 2009. Two-thirds of MINFOF and MINEP employees are expected to retire between 2005 and 2011; the new recruits will bolster staff quality and motivation without significantly increasing salary costs. A new capacity-building program will complement the recruiting strategy.

Notes

1. By December 1993, real GDP had declined by about 30 percent and real per capita income had dropped by 50 percent compared with the mid-1980s. The government faced severe liquidity constraints and was unable to meet its internal and external financial obligations.
2. This network includes 15 national parks (2,679,607 hectares), 6 forest reserves (702,995 hectares), 2 wildlife sanctuaries (67,000 hectares), 3 zoological and botanical gardens (4.07 hectares), 44 hunting zones (3,778,016 hectares), and 17 community hunting zones (1,163,724 hectares), which together occupy approximately 8,391,346 hectares.
3. Yale Center for Environmental Law and Policy and CIESIN (2005).
4. As chapter 3 indicates, little political will existed to support MINEF even after the economy began to recover. In 1995, the government supplied only about 19 percent of MINEF's small and externally supplemented budget (totaling $12.4 million). By 2007, the government was supplying 64 percent of the total budget—which at $30.3 million was more than double the 1995 level.

CHAPTER 2

The Advent of Forest Reform

Following the economic crisis of 1985, Cameroon urgently required international support to restore its macroeconomic equilibrium, its productive sectors, and its economic growth. The forest sector was not initially identified as a priority by the political elite. However, a 1988 assessment of the sector by the Tropical Forest Action Plan (TFAP) attracted considerable attention by stating that the forest sector's contribution to the national economy was vastly below potential, representing less than 2 percent of GDP, and that the forest industry was obsolete, wasteful, and ecologically damaging. At the same time, as already noted, Cameroon's forests were integral to its well-organized political patronage system, fueled by corruption and vested interests.

Forest sector reforms originated in this context. In hindsight it is clear that they were driven primarily by three forces: the 1994 Forest Law, economic leverage, and the synergies and partnerships that developed among reformers in government and civil society. These forces are discussed in the sections that follow.

The 1994 Forest Law

Enacted in 1982, Cameroon's forest code could no longer respond to the economic, social, and environmental imperatives emerging in the forest sector in the early 1990s. The same rules applied indiscriminately to all forest areas, regardless of the forest's specific importance to nearby communities, indigenous peoples, government, or industry. The rights of local people in particular were narrowly defined in terms of "user rights."

The code lacked any provision for sustainable forest management, and the short-term (maximum of five years) licensing arrangements encouraged quick extraction. Forest taxation was based on export duties. The potential for taxation to encourage sustainable practices was entirely untapped.

If Cameroon's forest sector was to fulfill its potential and help revitalize the economy, the legislative and regulatory framework for forest resource management clearly had to change. In 1993, the government of Cameroon adopted a forestry policy with well-defined, long-term objectives. The forestry policy was the basis for the draft Forestry Law of 1994, which was submitted for debate in the National Assembly. The draft law introduced five basic reforms:

1. The forest estate was classified into zones according to the highest-priority use: permanent forests, including protected areas and forests for commercial production, and nonpermanent forests (see figure 3.1 in chapter 3).
2. Long-term forest harvesting rights were allocated through a public auctioning process based on technical and financial criteria.
3. Government institutions were reorganized to focus on forest governance (regulation and control). Productive activities were transferred to community forests and production forest operations.
4. Private firms granted long-term concessions to operate in permanent commercial forests had to develop and implement forest management plans, which would be monitored by the Forest Administration.
5. Communities and local councils could establish their rights to manage forests through a contractual relationship with the administration.

The law was developed when the executive branch of the government was under severe financial constraints. The National Assembly, which did not see itself under the same duress, changed some of the law's key provisions. For example, the size of long-term forest management units (UFAs, also referred to as concessions) was limited to 200,000 hectares and their duration to 15 years, well shy of a sustainable harvesting cycle. Assembly members challenged the auctioning provision, arguing that national firms could not compete with foreign companies. Responsibility for forest management remained with the government, with "possible transfer" to the private sector at some later date. The exportation of logs was banned to promote local processing.[1] The World Bank regarded the resulting law as a step forward but not as satisfactory, given the lack of transparency introduced into auction procedures, the limitations imposed on the size of forest concessions, the excessive role of the state, and the potential perverse economic effects of the ban on log exports. At any rate, even with the adoption of two important implementing decrees in 1995, the law remained relatively inactive until 1998, owing to the lack of corresponding regulations, political commitment, and institutional capacity.

Economic Leverage

The economic crisis gave the World Bank and the International Monetary Fund an opportunity to introduce and support far-reaching reforms in the forest sector, which would encourage the sector to contribute to economic growth and strike at the heart of the patronage system. The forest sector was a focal point of three successive structural adjustment programs negotiated between the World Bank, the IMF, and Cameroon: the Economic Recovery Credit of 1994 and the second and third Structural Adjustment Credits—SAC II and SAC III—approved in 1996 and 1998, respectively (see appendix 1 for details). However, both the Economic Recovery Credit and SAC II enjoyed little government ownership and thus did little to further the reform agenda. Unlike the previous structural adjustment programs, SAC III contained detailed and specific conditions related to forests. These measures, discussed in detail in the next chapter, were designed to generate and test political commitment to the 1994 Forest Law, create a regulatory infrastructure to implement the Forest Law, design and enforce a new forest taxation system, improve transparency and governance, and combat corruption throughout the sector. Together, these measures would create a more transparent, organized, and sustainable system for accessing forest resources that would benefit more of Cameroon's people and the environment.

Synergies and Partnerships

The reforms would not have progressed without the collective energies and critical input of a range of partners.

International Monetary Fund

The IMF's focus on transparency and economic reform was instrumental in helping the government reform its forest sector. SAC III and related forest issues and reforms featured prominently in the IMF's Poverty Reduction and Growth Facility,[2] as well as in the monitoring of the HIPC

Initiative. The constant inquiring work from the IMF on significant issues of forest reform drew attention to the dialogue and ensured that the forest sector remained under the eyes of decision makers.

The Donor Community

The policy dialogue on forest reforms initially occurred between the World Bank and government, but the debate broadened considerably in the late 1990s as the donor community—which largely supported the direction of the forest reforms—came to question the sequencing and priority given to specific reform measures. For example, France and Canada considered that Bretton Woods institutions gave too much attention to taxation and to awarding concessions and too little attention to promoting forest management plans. Although sometimes contentious, the debates greatly enriched the content of the reforms and fostered the development of a harmonized approach to forest sector reform. The approach was formalized in a code of conduct signed in January 2006 by representatives of 13 partners, including international nongovernmental organizations (NGOs).[3] The signatories included many who have long been active in Cameroon's forest sector: Canada, Germany, France, the Netherlands, the United Kingdom, the European Union (EU), the African Development Bank, the World Bank, the Food and Agriculture Organization of the United Nations, the United Nations Development Programme (UNDP), World Wildlife Fund (WWF), the Netherlands Development Organization (SNV), and the International Union for the Conservation of Nature (IUCN). The code of conduct includes a matrix jointly adopted by the government of Cameroon and the donor community to assess progress in the forest sector.

International NGOs

The interest of NGOs in pursuing transparency and forest conservation objectives, as well as in the government's institutional weakness and need to shore up its international credibility, helped create specific niches for well-respected NGOs to support forest reform:

- Collaboration with the WRI, Global Witness (GW), and Resource Extraction Monitoring (REM) enabled the government and other partners to detect suspicious activity in concessions and parks and made inaction hard to justify. Independent observers reported on and were associated with the quality of auctions and forest control operations, and they helped to create a public register of forest and wildlife infractions. The GW, building on brief U.K.-sponsored missions in 2001, became the first independent observer for forest law enforcement, supported by the World Bank, the United Kingdom's Department for International Development (DFID), and the European Union (in fact, this independent monitoring enabled Cameroon to become eligible for HIPC debt relief). The GW was succeeded in 2005 by REM, which works with MINFOF law enforcement agents to conduct field investigations and specific studies of governance issues facing MINFOF and the private sector. The independent observer relies on good field investigation techniques and information from local informants, follows up on legal cases that are detected, and reports findings to an internal reading committee and directly to the minister. Results of the GW and REM field missions have been published on the Internet.

- The Global Forest Watch (GFW) program of the WRI supports Cameroon's forest monitoring efforts through the use of remote sensing and geographic information systems and maintains an online cartographic and statistical database for users and managers of forest resources as well as other constituencies interested in Cameroon's forests. The World Bank brokered a formal relationship between the government of Cameroon and GFW under SAC III, the third structural adjustment program.

- The WWF has been involved in most major forest policy and conservation initiatives, including the first Summit of Central African Heads of State on the Conservation and Sustainable Management of Tropical Forests (in Yaoundé in 1999), the development of the national strategy for biodiversity conservation, and the management of major protected areas.

- The Wildlife Conservation Society has built an extensive collaborative relationship with MINFOF and CAMRAIL (Cameroon's national railway system), providing support and technical assistance to protect biodiversity in national parks and to curb poaching.

- The IUCN helps to develop policies for conserving biodiversity and managing natural resources, wildlife, and bush meat. It also supports the participation of parliamentarians and civil society in policy development and implementation oversight.

- The Last Great Ape Organization (LAGA) challenges the corrupt practices that threaten biodiversity conservation by collaborating with MINFOF to fight commercial poaching and illegal trading in protected species. LAGA does not merely monitor infractions but is involved in all stages of the enforcement and application of Cameroon's Wildlife

Law, from field arrest operations to the prosecution of offenders in court. It also seeks to reinforce responsibility among commercial forest operations, whose activities could facilitate violation of the Wildlife Law.

Notes

1. As discussed later, the ban was partially revoked in 1999.
2. Forest reforms were systematically reported in the government of Cameroon's Letter of Intent to the IMF.
3. By signing the code of conduct, partners expressed their formal commitment to using the Forest and Environment Sector Program as the framework for harmonizing and aligning their assistance to the government of Cameroon in areas related to forests and the environment. The code enacts joint mechanisms for financing, monitoring, and evaluating, among other activities.

CHAPTER 3

Objectives and Content of the Reforms

The overarching objective of the reforms was to replace chaotic and opaque arrangements for accessing forest resources, which benefited the few, with a more organized, transparent, and sustainable system that would benefit greater numbers of people and the environment. As unsatisfactory as the 1994 Forest Law may have appeared, it provided a nationally ratified platform for initiating forest reforms to support that objective. SAC III, approved in 1998, contained measures designed to refine and build upon the intent of the 1994 law in six strategic areas (table 3.1):

1. *Rationalizing the use of forest land* and setting the stage to secure rights on forest land, regulate access to forest resources, reinforce local rights to forest resources, expand protected areas, and conserve biodiversity.
2. *Enabling communities to benefit more significantly (especially economically) from their rights to use forest resources* by involving communities in conserving those resources over the long term.
3. *Allocating harvesting rights more efficiently and transparently* by linking them to responsible forest management and to tax revenues.
4. *Ensuring that forest management is socially and environmentally responsible* and that the forest resource will exist in perpetuity.
5. *Adopting a new forest taxation system* to promote sustainable forest management, share forest rents more equitably through greater

Table 3.1. Forest reform measures contained in SAC III

Area of reform	1994 Forest Law	SAC III forest measures
Forest zoning	Classification of the forest estate into permanent production forests, protected areas, and rural areas.	Adoption of a national strategy for forest concession planning, taking into account requirements for sustainable forest management.
Community forestry	Establishment of rights for local communities to manage community or local council forests through a contractual relationship with the administration.	Adoption of a right for local communities to preempt neighboring forests from being earmarked as ventes de coupe (where logging is permitted on a maximum of 2,500 hectares for a maximum of three years).
Title allocation system	Allocation of short- and long-term forest harvesting rights through a public auction in which technical and financial criteria are considered.	Adoption of new, detailed regulations for awarding harvesting rights, including revised selection criteria and an external validation mechanism involving an independent observer.
Sustainable forest management	Introduction of forest management plans implemented by private firms in permanent production forests and monitored by the Forest Administration.	Adoption of procedures to prepare, approve, and monitor forest management plans. Selection of international nongovernmental organizations to monitor and assess the implementation of forest management plans on the ground. Implementation of a guarantee system to ensure compliance with forest management plans.
Forest taxation	Mention of a system to redistribute a portion of the area tax to local councils and communities.	Adoption of reforms in forest taxation, including the creation of a program to enhance forest tax revenue (through better monitoring and recovery of forest taxes) and a system for redistributing annual area revenues (the state to receive 50 percent, local councils 40 percent, and local communities 10 percent).
Adaptation of the institutional framework	Reorganization of public institutions to focus on governance (regulation and control). Transfer of productive activities to the private sector, local communities, and local councils.	Revision of the institutional status of the National Forestry Development Agency to clarify its mandate and sources of financing.

Source: Authors.

community participation, increase local processing, facilitate tax collection, and improve governance and transparency in the forest sector.

6. *Reorganizing public institutions* to align forest sector organization and expenditures with policy objectives, expand accountability, and ensure that the state performs its essential functions and removes itself from others.

For the next 10 years, these six strategic areas became the focus of government collaboration with the World Bank and other partners. As new information became available, as perspectives changed, and as new concerns emerged—for example, concerns over climate change—they were reflected in the reform agenda. The sections that follow highlight some of the issues arising as the reforms were implemented. Implementation did not always take a clear, predictable path. The sequencing was not impeccable. Yet the process moved forward.

Organizing the Forest Landscape

Classification of the national forest estate into distinct zones (based on surveys of forest characteristics, population density, and use) provided a foundation for protecting forest land and also for specifying those forest areas that could be converted to other uses. Geographic zoning *identifies* which forest land belongs to which classification; the subsequent gazetting process *legally ratifies* the classification. Together, zoning and gazetting enable government, communities, industry, and other stakeholders to establish secure use rights. In Cameroon these actions made it possible to introduce some degree of regulation and clarity into a previously chaotic system under which the government acted as the landlord of the forest, while lacking the capacity to monitor and control its use. Zoning and gazetting also permitted some recognition of traditional forest use rights and made it possible to develop a national strategy for allocating forest harvesting rights.

The Zoning Process

Figure 3.1 depicts the hierarchy of forest zones in Cameroon, defined according to the 1994 Forest Law as permanent forests (including protected areas and production forests) as well as nonpermanent forests (including "rural areas," which are forested lands located near villages and settlements; in figure 3.1, they are in the "Other forests" category).

Figure 3.1. Cameroon's forest zones

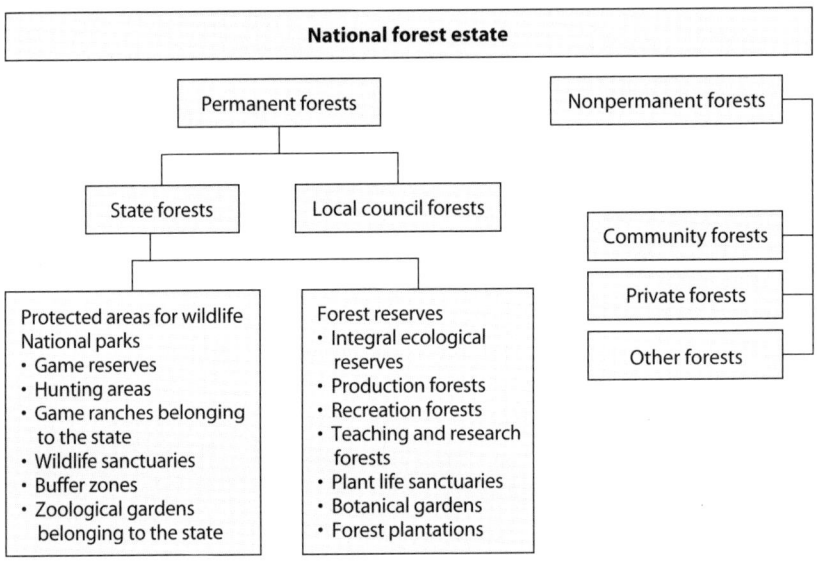

Source: GFW 2005.

According to the 1994 Forest Law, the *permanent forest domain* is to be extended gradually through gazetting to cover 30 percent of national territory, represent the full range of forest ecologies, and be managed sustainably. In this domain, local uses of forest resources are restricted, and production forests and local council forests occupy the largest share of the area. The *nonpermanent forest domain*—community forests, private forests, and other forests—is zoned for other uses and occupies about 5 million hectares. These areas may be converted to nonforest uses. Private forests have never been inventoried, and their area is not known.

Zones were identified geographically by remote sensing, based on vegetative cover and coarse demographic analyses. Local consultations, social studies, and assessments of traditional tree and land tenure arrangements were extremely limited and clearly insufficient. In some instances the government appeared to think that a zoning map could simply substitute for consultation. As a result, some observers have regarded the zoning plan as an instrument to remove forests from the control of local populations. In fact, zoning and especially gazetting have helped to clarify forest use rights and supported communities' ability to challenge regressive practices and prevent infringements on their rights (box 3.1).

Box 3.1

Did forest zoning remove land from local populations?

Traditional rights to local forest land and products may have endured in Cameroon for centuries, but they were not recognized in any official sense. Since colonial times, all forest land had been vested in the state. Zoning and gazetting were undertaken specifically to remedy this lack of secure use rights. Forest zoning and gazetting first determine how specific areas of forest land would be used (or not used) and then clarify and enforce the rule of law with respect to those decisions. With clearly defined use rights, regressive colonial practices and relations of power can be challenged. Citizens can express their views and take legal action to challenge infringements on their rights.

In Cameroon, it is forest gazetting—not zoning—that confers these legal powers. Gazetting places forests in a given legal category, and land ownership can be formalized only after title to a given area has been debated publicly and opposing claims heard. Without question, forest zoning and gazetting in Cameroon could have been more effective, democratic, and participatory, but data on gazetting show that boundaries proposed in Cameroon's original zoning plan frequently have been modified to take the current occupation of land into account, often to the detriment of timber concessions that have already been allocated.

Source: Tchuitcham 2006; authors

The forest zoning plan is expected to cover the totality of the country's forests (estimated at 30 million hectares). The plan has been completed for southern Cameroon, where 14 million hectares have been zoned. Of these, about 6 million hectares have been apportioned as rural community and private forests, about 6 million for long-term forest production (forest concessions), and the balance for conserving biodiversity.

The Gazetting Process

Gazetting is a highly sensitive and potentially controversial process, given that it legally establishes the status and prospective use of a specific forest area (figure 3.2). Gazetting entails consultation with local communities, a process that—to compensate for limitations pointed out earlier—can dramatically alter the boundaries proposed by the

Figure 3.2. The gazetting process

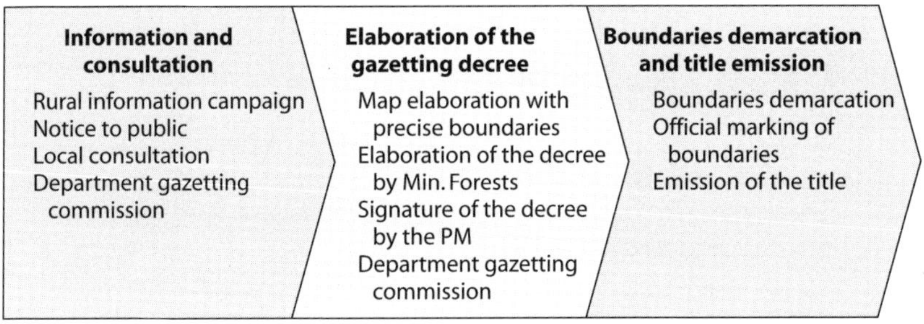

Information and consultation	Elaboration of the gazetting decree	Boundaries demarcation and title emission
Rural information campaign Notice to public Local consultation Department gazetting commission	Map elaboration with precise boundaries Elaboration of the decree by Min. Forests Signature of the decree by the PM Department gazetting commission	Boundaries demarcation Official marking of boundaries Emission of the title

Source: Authors.

zoning plan (box 3.2). By promising to clarify the land tenure rights and obligations of the state and its citizens, which some governing bodies resist, gazetting has undeniable political implications. The delays experienced in obtaining the prime minister's approval for the first group of forest lands proposed for gazetting may have resulted from the novelty of the approach, the sheer number of transactions, and the vast amount of land under consideration. Not surprisingly, some draft gazetting decrees languished, unsigned, in the prime minister's office for more than two years.

The creation of UFAs, which are areas of production forest deemed suitable for sustainable use under a 30-year managed harvesting cycle, was an essential step in implementing the forest management policy introduced by the 1994 Forest Law. Under pressure from its external partners, including the World Bank, in January 2000, the Forest Administration gave serious attention to establishing UFAs, starting with those allocated in 1997. By 2006, among 96 UFAs awarded, 45 had been formally gazetted (appendix 2). These 45 units accounted for 51 percent of the area allocated to production forest, covering more than 3 million hectares.

Protected areas were also zoned and gazetted following the same process as UFAs. The rainforest area of Cameroon today has 3.9 million hectares of national parks, wildlife sanctuaries, wildlife reserves, and other conservation areas. Additionally, about 900,000 hectares of UFAs, initially intended for logging in the zoning plan, have not been allocated to industry but reserved for conservation in the national zoning plan (appendix 3).

Box 3.2

Impact of consultations on size and boundaries of forest concessions

In most consultations related to the gazetting of production forests, local populations claimed rights to land that the zoning plan had slated for new timber concessions and requested that the proposed boundaries be adjusted. Despite initial resistance from authorities, the boundaries of most concessions (current and proposed) have been adjusted to reflect local land claims.

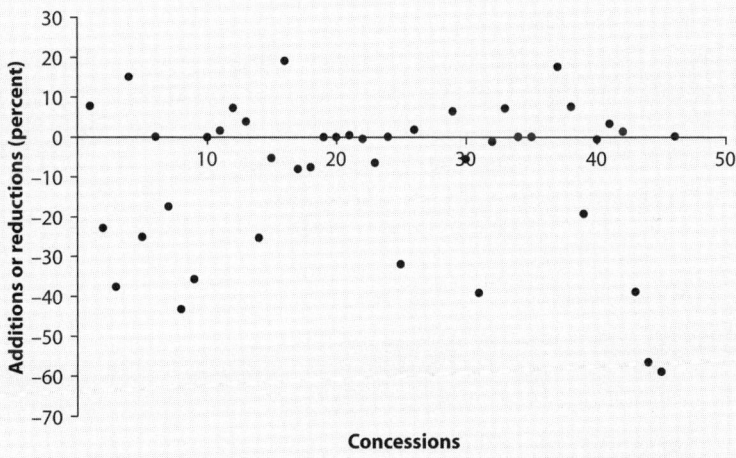

The figure illustrates the variation in area after gazetting is completed. Although no specific studies have been done to ascertain the extent to which legitimate claims were taken into account, it is noticeable that boundaries were adjusted for 36 out of 45 zoning proposals. Adjustments ranged from a reduction in the proposed area of 59 percent to an increase of 19 percent. When possible, the administration tried to compensate for the reduced area by dividing unallocated forest management units or mining areas, which in some cases increased the area of concessions that had been reduced.

Source: Authors, based on data from MINFOF.

Efforts to Foster Community-based Forest Management

Community forestry was introduced in the humid tropics within the last 25 years, initially in Asia. It has experienced mixed results stemming from variation in local factors such as management traditions, the relationships between forest users with respect to their different land tenure rights and powers, the existence of markets for certain products, the cost of accessing those markets, the relationship with local governments and authorities, and the behavior of the administration. Among all of the innovations introduced in Cameroon's forest sector by the reforms, community forest management arguably remains the most challenging.

Before adoption of the 1994 forest law, there was no concept of community forestry in Cameroon's forest regime. The law sought to engage communities in managing forests sustainably and conserving biodiversity, while at the same time reducing rural poverty and improving governance in the forest sector. New regulatory and institutional frameworks developed under SAC III made it possible to establish community forests (on village forest land within the nonpermanent forest domain) and local council forests (on state-owned forest land).

Community Forests

Throughout Africa's humid tropics, the term *community forestry* refers to organized forest production, mostly of timber products. In Cameroon, the term *community forestry* refers specifically to a forest placed under the management of a community at its behest because the community has customary rights to that forest. It is important to emphasize that as is the case with forest concessions, community forests do not involve the transfer of land ownership, but rather the transfer of the right to manage and use the forest land.

Community forests are actually a subset of the forests used by communities—a distinction not always maintained in discussions about community forests. Most of the forest products obtained by rural and indigenous populations, either for trade or their own consumption, do not come exclusively from community forests but from the much broader forest space to which rural and indigenous people have unchallenged legal or de facto access, including forest concessions, community hunting zones, and some protected areas. The regulation of forest-based income-generating activities also extends well beyond what is normally considered the community forestry framework. The importance of community forests can therefore not be measured by simply comparing the area under community forests with the area under national parks and production forests.

Cameroon's first community forests, Mbimboé and Bengbis, were established by 1997, but further progress was slow. To speed the adoption of community forestry, the government was given a specific deadline to identify prospective areas for designation as community forests. The conceptual dissonance of this process—asking government to dictate the location of a community forest in a domain (the nonpermanent forest) where government was not supposed to intervene—called for an adjustment. In December 2001, communities were officially granted the right to stop the allocation of a vente de coupe in their locality by claiming their interest in creating a community forest, which they would manage themselves. Communities very frequently invoke this preemptive right (*droit de préemption*); they then have two years to make a final decision and prepare to manage the community forest area that has been set aside for them.

Every community forest must be managed under an agreement signed by the community and the local Forest Administration office. Under this contract, the Forest Administration entrusts a section of forest (no more than 5,000 hectares) from the nonpermanent national domain to the local community, with the understanding that it will be managed, conserved, and otherwise used in the interests of the community (Oyono 2005). The community is entitled to technical support from the Forest Administration to manage the forest in accordance with a simple management plan defining the activities to be undertaken (Oyono 2005).

With very few exceptions (see box 3.3), the search for revenue has been the impetus for creating community forests. According to MINFOF,

Box 3.3

Protecting forest biodiversity at the urban margin: Bimbia Bonadikombo Community Forest

The rainforest on the western foothills of Mount Cameroon has many species of national and international importance. Its vibrant biodiversity—1,500 species, of which 24 are endemic and 43 rare or new to science, are gathered in a relatively small area—is under threat, given the area's proximity to the town of Limbé (population 85,000) and its surrounding settlements. Economic hardship and population growth have heightened the demand for forest products.

Box continues on next page

Box 3.3—continued

In 1998, local people, concerned about the dwindling forest, joined forces with the Mount Cameroon Project, the local Forest Administration, and other stakeholders to create the Bimbia Bonadikombo Natural Resource Management Council (BBNRMC) and establish a community forest to protect and manage the remaining forest. Plans for Bimbia Bonadikombo Community Forest were developed and approved; by August 2002, the BBNRMC was actively managing the forest.

The forest occupies 3,735 hectares. Most of it—especially the northern part—has been relatively degraded as a result of human activity, including subsistence farming, bush-meat hunting, fuelwood collection, and artisanal logging. The biodiversity-rich southern part of the forest (1,229 hectares) is set aside for conservation, and all extractive activities are forbidden. Volunteers delegated by village chiefs regularly patrol the area. In 2006, income from fines and sales of confiscated products amounted to about $3,680 (CFAF 1,840,000), representing about 45 percent of total income from permits issued for use of the forest. The BBNRMC is encouraging ecotourism and research, hoping that such activities will increase the benefits derived from conserving the forest.

Conservation in the south is complemented by efforts to use the northern forest sustainably. Unless the northern forest areas can offer alternative livelihood options, the overall conservation strategy will fail. The BBNRMC grants forest use permits to different groups—small-scale timber cutters, charcoal burners, and fuelwood collectors, among others—based on predetermined quotas (unfortunately set in the absence of a forest inventory). Given that agriculture is the foremost threat to the forest, BBNRMC supports the adoption of agroforestry to improve yields and discourage farmers from clearing new land. Three nurseries have been built with technical and financial support from the World Agroforestry Centre.

Clearly, management of the Bimbia Bonadikombo Community Forest occurs in a complex setting—the margin of urban development, where biodiversity conservation and livelihood issues intersect in unusual ways. The current management approach has shown some positive impact on forest conservation, but it is increasingly obvious that these local efforts simply cannot continue for long without external resources and partners. Given the still-limited number of tourists in the area and the extreme difficulty of resisting demand for fuelwood from Limbé, it is feared that this model is not sustainable.

Source: Authors; Godwin and Tekwe 1998.

as of April 2008, 177 communities had gained approval for management plans covering nearly 632,330 hectares from the Forest Administration (figure 3.3). By 2008, agreements had been concluded to create 143 forests on 545,944 hectares. More than 75 percent of community forests are located in Eastern, Central, and Southern Provinces, forest-rich areas where communities are familiar with logging.

Local Council Forests

The 1994 Forest Law permits a local council to create its own forest "estate" (*Domaine privé de la Commune*) in the permanent forest domain, on the condition that it prepare a management plan that is approved by the Forest Administration.

Like community forests, local council forests have been promoted by external donors. Perhaps the best-known example is the first local council forest—the Forêt Communale de Dimako, a 16,000-hectare forest approved in 2001 on the basis of a management plan developed with financing from the French government. Currently, only six local council forests are gazetted (Dimako, Djoum, Gari Gombo, Messondo, Mouloundou, and Yokadouma), totaling about 153,500 hectares, of which 121,000 hectares are managed according to an approved plan. Another 11 forests (nearly 247,000 hectares) are at various stages of gazetting, and a further 4 (nearly 92,000 hectares) are preparing to initiate gazetting (Ngueneng and others 2007).

Figure 3.3. Creation of community forests as of April 2008, Cameroon

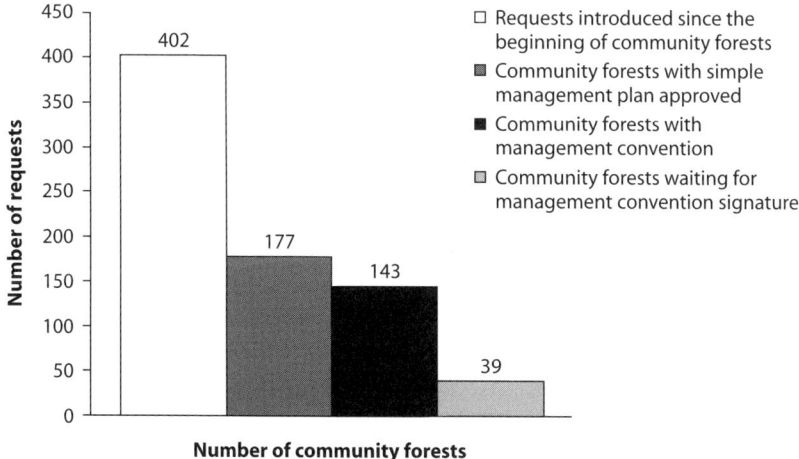

Source: MINFOF, 2008

Local council forests appear to provide significant revenue to decentralized local administrations, in addition to the 40 percent of the annual area tax that they receive if their land overlaps with UFA land (an opportunity that mayors seek to exploit). The first national congress on local council forests, held in Yaoundé in June 2006, was attended by mayors from forest areas to exchange experiences and build on lessons learned.

A New Strategy and System to Allocate Forest Harvesting Rights

Until the late 1990s, rights were awarded to harvest relatively small forested areas or volumes of timber under short-term contracts. These contracts did not appear to be awarded in a rational or orderly way that reflected the sustainable management goals of the 1994 Forest Law. At that time it was not unusual for more than one person or company to be given the right to harvest the same area or for harvesting rights to be granted within protected areas. To rectify these problems, in 1999 the government adopted a planning strategy and disseminated a booklet that summarized the features of all harvesting rights and indicated how they would evolve over the coming three years.

Strategy to Reorganize the Production Forest Estate

Harvesting rights defined under the 1994 Forest Law included mainly the long-term rights to UFAs, short-term ventes de coupe, and small titles (30-cubic-meter personal authorizations, 300-cubic-meter cutting permits, and the recovery of timber following land conversion or special authorization). Ventes de coupe—which are established in the rural domain, in nonpermanent forest that can be converted to nonforest uses if local communities so decide—permit harvesting on a maximum of 2,500 hectares for a maximum of three years.

Through the booklet on planning strategy, the government disseminated its plan for sustainably managed UFAs to replace licenses issued before 1994 and gradually become the main source of commercial timber for Cameroon. All volume-based harvesting rights, such as ventes de coupe and small titles, would decrease over time. Although small titles continue to exist (and serve as an entry point for illegal logging), by 2006 forest concessions operating under management plans supplied more than 85 percent of commercial timber, up from less than 30 percent in 1998.

While primarily directed at industry, the strategy and booklet reaffirmed the intent to develop community forests. They also stated the government's intention to reduce total harvests to bring logging down to sustainable levels. Most important of all, the strategy was accompanied by maps that showed valid titles and made it easy to unmask logging operations based on expired or illegal titles. The booklet and maps were updated several times and made available to interested parties.

The planning strategy made it possible to predict which concessions or ventes de coupe would become available and helped the private sector prepare for possible participation in competitive allocation. The strategy centralized the planning, allocation, and monitoring of concessions and forestalled discretionary allocations by provincial and departmental authorities—allocations that had been much more difficult to detect and eradicate.

The Concession and Vente de Coupe Allocation System

The reforms introduced a competitive bidding system for awarding 15-year renewable harvesting right to forests, with provisions for sustainable forest management. To secure the winning bid for a concession, a bidder must demonstrate adequate technical expertise and investment capacity and offer the highest financial bid. In acquiring a concession, the winner commits to develop a management plan and to comply with specific technical, social, and environmental obligations. Cameroon was the first country in the world to adopt this approach. Comparable approaches are being adopted in other countries in the Congo Basin and Ghana and are being considered in Brazil.

Less than 10 years after this system was implemented in Cameroon, harvesting rights went from being administratively distributed, short-term concessions, free from any forest management obligations, to being long-term, competitively awarded concessions, accompanied by forest management obligations spanning the duration of the contract.

Competition rapidly increased the rental of forest land (in the form of an annual tax, based on the area of the concession). From a base of $0.60 per hectare in 1990 (set administratively), the area tax rose to averages of $5.60 per hectare for UFAs in 2006 and to $13.70 per hectare for ventes de coupe in 2005. Perhaps most important, the bidding process attracted more environmentally and fiscally responsible companies to Cameroon.

How the Auction Works

The allocation of UFA concessions and ventes de coupe is supervised by the Interministerial Commission for Awarding Forest Concessions (Commission Interministérielle d'Attribution des Concessions Forestières), which includes representatives of various ministries (Forests, Finance), representatives of trade unions, individual experts, and an independent observer. The independent observer is a private individual from outside the government who attends meetings of the Interministerial Commission, reviews bidding documents, observes the process for making award recommendations, and prepares a report for the Minister of Forests on the integrity of the process.

The selection criteria and procedures are governed by regulations that are periodically revised in light of lessons learned from successive rounds of auctions. Bidding rounds are preceded by public sessions to explain the system to potential bidders. Any company registered in Cameroon may compete.[1] Detailed instructions are provided to bidders in the bidding documents, and the procedures for evaluating bids are detailed in published official acts. The selection criteria are financial and technical (including the firm's technical capacity, its prior industrial investments and financial strength, and its respect for fiscal and environmental commitments). The financial offer is based on the value of the area tax offered per hectare and per year, in excess of the floor value prescribed by the Finance Law.[2] (For additional discussion of the merits of the bidding system, see appendix 4.)

Companies awarded UFA concessions have 45 days to provide an advance guarantee bond (*lettre de garantie*), which is equivalent to one full year of area tax. Companies failing to pay by the agreed date are excluded, and concessions are awarded to the second-highest bidder.

For a discussion of how the area fee could influence loggers' behavior, see box 3.4.

Performance of the Auction System

A quantitative measure of the performance of the auction system is the value of the annual area tax, which was previously set by the administration but became subject to bidding after the auction system was introduced. The area tax has increased substantially since the auction system came into force. While other factors such as the international market and forest quality certainly affect the bids that are offered, there is no doubt that implementation of the auction system initiated the rise in bids for harvesting rights.[3]

The efficiency of the auction system depends to a great extent on ensuring genuine competition and confidentiality. Competition and confidentiality also ensure that bids truly express willingness to pay. Sustained performance depends on frequently updating rules and processes

Box 3.4

Does the area fee influence loggers' behavior?

Boscolo and Vincent (2007) have suggested that "area fees [the annual tax based on the area of the concession] can induce concessionaires to accelerate timber harvests and to harvest more selectively" and add that "in Cameroon, area fees at recent levels create an incentive for concessionaires to harvest forests in half the estimated sustained-yield period."

Numerous motivations exist for violating the annual allowable cut, especially when concessions do not occupy vast areas, as they do in other Congo Basin countries such as the Republic of Congo and Gabon. For instance, high-speed logging is typical of most Asian loggers, whether they operate in Southeast Asia or Africa, and regardless of the fiscal structure imposed by a given country. Their "business model" relies on quickly harvesting large volumes and on ensuring high mobility of the invested capital (conglomerates pursue not only forestry but various agribusinesses, such as oil palm harvesting and processing, in various locations). Perhaps the greatest incentive for quick forest depletion is the private discount rate. Central Africa is a region with high institutional and political risks, and businesses usually confess they invest only if they can recoup their investment within a couple of years. In Central Africa the desire to see a rapid return on investment tends to increase the speed of the harvest, not the intensity, which is limited by the floristic composition of the forests as well as market characteristics. Thus it would seem that the high private discount rate would explain loggers' behavior more than the forest tax structure.

Boscolo and Vincent have observed that prospects for replicating Cameroon's success with area fees in other countries "is not clear, as the measures were conditionalities attached to a series of structural adjustment loans from the World Bank and the International Monetary Fund to the country." Their recommendation, which reflects steps taken in Cameroon, is that "countries that wish to encourage concessionaires to comply with sustained-yield requirements must implement measures that counter the depletion-accelerating effects of area fees." Several such measures have been implemented in Cameroon, and others that could potentially reinforce the current tax framework are detailed in this report.

to eliminate gaps and close loopholes that inevitably emerge as institutions and bidders gain experience with the system. After five years of
relatively good performance, signs of weakening were seen in the level of
confidentiality and in the oversight of the independent observer, whose
appointment remained unchanged over several auction sessions, despite
regulations that mandate otherwise and are designed to prevent collusion. As a result, the auctions of 2005 and 2006 saw reduced offers
(figure 3.4), suspicious numbers of bidders eliminated at the technical
evaluation stage, and harvesting rights awarded to the only remaining bidder at the floor price. This situation led the World Bank to ask the govern-

Figure 3.4a. Evolution of the Area Tax

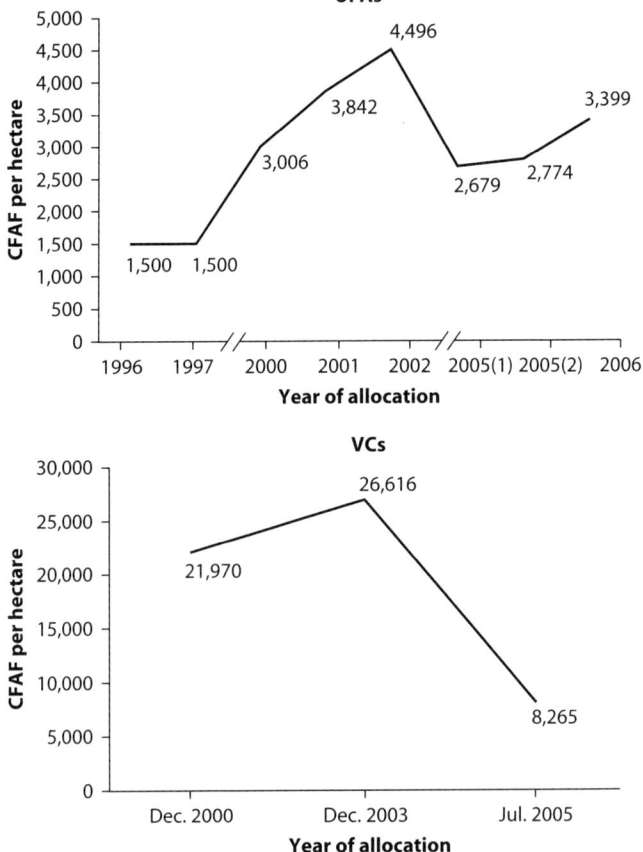

ment to investigate weaknesses in the system and adopt improvements before the next auction session. The introduction of these improvements is pending.

Reforms to Support Sustainable Forest Management

The central consideration in devising Cameroon's forest management prescriptions was to move the timber industry toward more environmentally and socially sustainable practices. To understand the effects of these prescriptions, it is helpful to understand their elements.

Figure 3.4b. Evolution of the Area Tax—continued

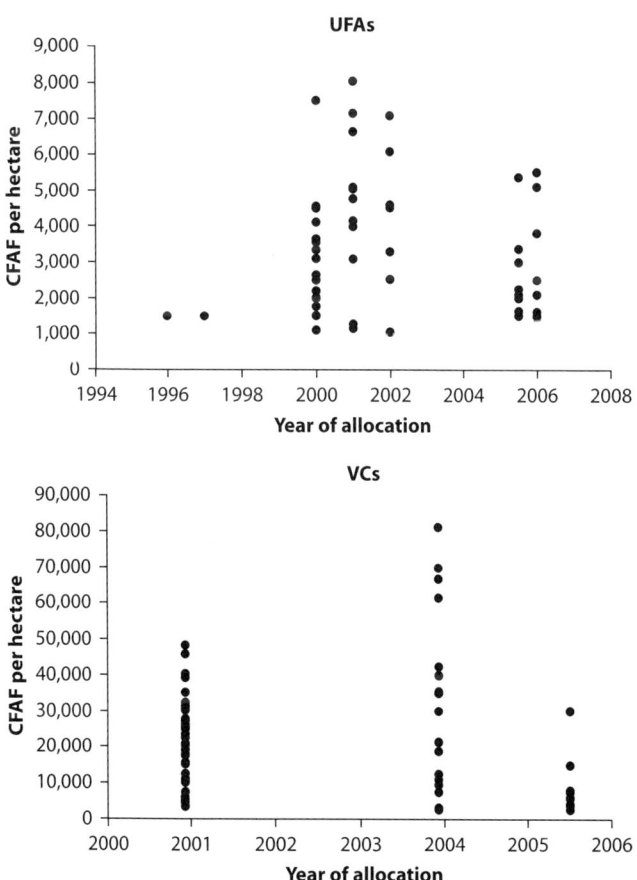

Source: Authors.

Managing Cameroon's Forest Production: The Basics

A production forest management plan begins with topographical, biological, and socioeconomic surveys. The survey data guide decisions on dividing the concession into conservation series (relatively small areas within the concession that have exceptional biodiversity values and are closed to logging), agroforestry series (areas where farmers and other sedentary populations may continue using the land for farming and other activities), and forest production series (where industrial forestry is practiced).

The forest production series is divided into a number of subareas (annual coupes). The number of annual coupes is based on the interval between harvests—the harvesting cycle. The length of the harvesting cycle depends on the time required for the annual coupe to recover from being harvested. In this interval, the remaining, smaller trees must reach the minimum permissible size for harvesting, and a new generation of desirable tree species must become fully established. Harvesting cycles usually range from 25 to 50 years. In Cameroon they have been set at 30 years.

The size of each annual coupe is determined in ways that minimize differences in annual production. The annual coupe will be smaller in tree-rich areas and relatively larger in areas with fewer trees. Once harvested, each annual coupe is left to rest for the duration of the harvesting cycle, and access tracks to that particular coupe are destroyed or abandoned.

Selective harvesting involves felling a few commercial trees per hectare every 30 years. When cutting a given species, foresters must ascertain that enough trees of that species of adequate size remain for that species to regenerate and maintain its long-term presence in the local ecosystem. To foster this outcome in Cameroon and preserve the economic value of the UFA, forest management plans impose tough constraints on harvesting most high-value species (see the next section). These requirements heighten the incentive for companies to harvest more "secondary" species, even if they are less valuable per unit of volume.

Of the more than 100 species generally available in the tropical humid forests of Central Africa, fewer than 12 are usually harvested by any individual forest company. Well-managed, selective harvesting seeks to alter the composition and the biodiversity of the forest as little as possible, but some alterations cannot be avoided (box 3.5). It is essential to bear in mind that maintaining forests in their original state is what protected areas and biodiversity reserves are designed to do, and this is why it is vital for them to be established in advance of production forests.

Box 3.5

Regulated selective harvesting: Diversity and other considerations

Do selectively harvested production forests maintain the diversity of tree species they contain? The ratio of species is likely to change over the long term. Because the standing volume that is first harvested from a primary forest "has accumulated over a long period, the commercial timber is likely to be of a quality and volume that will not be matched in future cuts . . . unless the logged forest is closed to further exploitation for a century or more. In this sense the first crop is, in practical terms, not repeatable" (Poore and others 1989). Light-loving tree species will be favored by the opening of stands, but selective harvesting may not open the forest canopy sufficiently to favor those species (Karsenty and Gourlet-Fleury 2006; a similar observation on the forest canopy was made for the Amazon by Fredericksen and Putz 2003). Species diversity may increase temporarily as a result of these changes, but it could also diminish at the regional level as species within production forests become more homogeneous (Bawa and Seidler 1998).

   Are there additional medium- and long-term issues to consider? Industrial forest management strategies are based on the expectation that the number of species accepted by the market would increase in the 30 years between the first and next harvesting cycle. To maintain the profitability of harvests, industry relies more on finding an increased number of marketable species in the next harvesting cycle than on waiting for the same few species harvested in the first harvesting cycle to recover their original number and volume. Although this strategy is supported by respected forest stewardship agencies, the choice of the medium over the longer term can be debated ad infinitum because data on the regeneration of commercial species are insufficient and cannot lead to informed decisions. Evidence for periods longer than a century would be needed.

   Could highly commercial species be planted instead of being left to regenerate? Natural regeneration is the norm in tropical production forests. Public forest managers and other experts do not regard enrichment planting as financially viable in Cameroon's UFAs. The UFAs are very large, and new trees would have to be planted in the highly dispersed locations from which the old ones were harvested.

Box continues on next page

Box 3.5—continued

Will industry start to harvest more "secondary" species? Yes, but the timing cannot be predicted easily. Timber markets have so far shown themselves to be conservative and slow in accepting unfamiliar but technologically fine secondary species. The problem is compounded when harvesting costs are high, because timber companies tend to become even more selective, concentrating on the most economically rewarding species. To facilitate the industry's transition to more secondary species in Cameroon, the tax burden should be alleviated for species that are particularly resilient and abundant. Adopting reduced impact logging (RIL) techniques would help mitigate the damage caused by increasing the harvest intensity. Note that many companies implement RIL as a prerequisite for third-party certification.

Can long-term forest management substitute for preserving forest biodiversity in national parks and other reserves? The forest management strategy adopted in Cameroon ideally will make production forests more stable, productive, and less subject to degradation. These forests will take some of the pressure off of protected areas and also provide essential environmental services, but they cannot replace national parks and other areas designed to preserve pristine rainforests.

Implementing Forest Management Plans in Cameroon

In Cameroon, production forests are part of the permanent forest estate. They cannot be converted into plantation forests and must be operated under long-term management plans. These plans are essential for sustaining production forests, given that under previous arrangements many of them were harvested at least once, especially in areas near rural settlements or served by roads.

The scope and content of management plans, as well as measures for monitoring their implementation, are clearly defined in a brief (18-page) decree (MINEF 222/2002). These legally binding guidelines summarize complex norms previously dispersed in hundreds if not thousands of pages of technical documents, guidelines, and regulations. The guidelines have considerably simplified the relationship between the Forest Administration and the private sector.

This reform represents a breakthrough in resource management, both because it introduces social and technical obligations and because the management of a publicly owned resource is delegated to the private sector.

All holders of harvesting rights to UFAs must design and implement forest management plans.[4] This policy marks a departure from timber-harvesting patterns determined solely by the market to a more organized, socially and environmentally responsible, long-term approach to forest management. Management plans must be completed within three years of the concession award; during this time the concession holder retains only provisional rights and can use only a small portion of the concession. Rights are canceled if the holder fails to prepare an approved management plan. Management plans are analyzed by a technical subcommission and then submitted to an interministerial committee that meets at least twice a year.

Management plans must contain a description of the UFA's natural and social environment, cartographic data, a forest management inventory, a definition of zones and user rights, the marking of UFA boundaries, and a calculation of the UFA's timber production potential. Important mandatory requirements are that traditional rights must be surveyed and respected by the concession holder and that socioeconomic surveys and consultation must be used to define spatial organization and secure customary user rights within a UFA.

From a technical perspective, the key parameters to be adhered to are the area and boundaries where timber can be harvested legally each year and the minimum diameter of trees that can be harvested legally (box 3.6). To determine these parameters, the company must complete an inventory and process the data using an algorithm that simulates forest growth in subsequent harvesting cycles. The simplicity and limited number of parameters to be checked by the Forest Administration are crucial to the feasibility of control, given the government's limited capacity.

From a company perspective, the preparation of a management plan represents an additional investment estimated at $5–12 (CFAF 3,000–7,000) per hectare, most of which goes to pay for the forest inventory, technical expertise, and data gathering. Management plans can profoundly alter the costs and benefits of timber enterprises. Management plan prescriptions limit the area open for harvest each year and may increase the minimum diameter of harvestable trees, thereby reducing benefits. On the other hand, the improved knowledge of forest resources afforded by management plans allows less costly and more efficient road networks to be developed and informs the company about secondary species that can gainfully enlarge the range of marketed species. Whether these opposite impacts result in a net cost savings or cost increase depends

Box 3.6

Ground rules for harvesting timber from Cameroon's forest management units

The ground rules for harvesting timber from Cameroon's forest management units are intended to foster *the selective harvest of a few trees per hectare once every 30 years within a given UFA.*

How much area is harvested? On average, only 1/30th of the area can be harvested in a given year (the annual coupe). There are some provisions to minimize damage to the remaining stand during the harvest. The same annual coupe cannot be harvested again for 30 years. In the intervening years, the forest is left to regenerate naturally.

Which trees are harvested? Concessionaires may harvest only trees that have reached the minimum diameter permissible for cutting (MDC), which differs from species to species. Minimum diameters were initially set administratively, but they must be revised (and are generally raised) based on the species distribution profile revealed by detailed inventories and the formula used to predict harvestable volumes for the second harvesting cycle. Trees to be harvested must be mapped and marked for records.

Although all species that have reached their MDC can be harvested, the concessionaire is interested mainly in harvesting the species for which there is a market and whose selling price compares favorably with production costs. There are provisions to ensure that sufficient intermediate-size trees of the most valuable species (trees that do not reach the minimum diameter) are left for the subsequent harvesting cycle.

What happens between the first and second harvest? Usually more trees of the more valuable species are harvested in the first harvesting cycle than in subsequent harvesting cycles, simply because the trees have had a longer time to mature. Aside from time, other variables affect the rapidity with which forests regenerate and the composition of tree species. For example, harvesting larger trees destroys some smaller trees and seedlings, but opening

on the intrinsic qualities of the forest, the cost structure and cost elasticity of the timber operation, the firm's forest management and marketing strategy, and its processing performance. In general, the impact on cash flow is negative in the short term, because harvests of primary species are reduced, and positive in the long term, because sustainable timber the canopy boosts the growth of light-demanding species. The intensity of the forest canopy opening affects the composition of species in the regenerated forest, depending on whether the remaining species are better adapted to light or shade. Many—but not all—commercial species are light demanding.

What happens during the second harvest? After 30 years, the same 1/30th of the concession may be harvested. The trees of intermediate diameter left during the first harvest will be suitable for harvesting. By that time, more species may have become marketable. If this is the case, subsequent harvests can also garner very large trees. Even so, volumes of the most harvested trees are generally lower in the second felling, because the initial harvest is composed of overmature trees (with diameters up to 200–250 centimeters), a volume that cannot be recovered within a 30-year harvesting cycle. Under law, management norms require a minimum of 50 percent of volume for the group called "managed species" to be recovered from one harvesting cycle to the next, and MDCs are raised accordingly to reach this threshold.

The figure provides a very simplified example of this process.

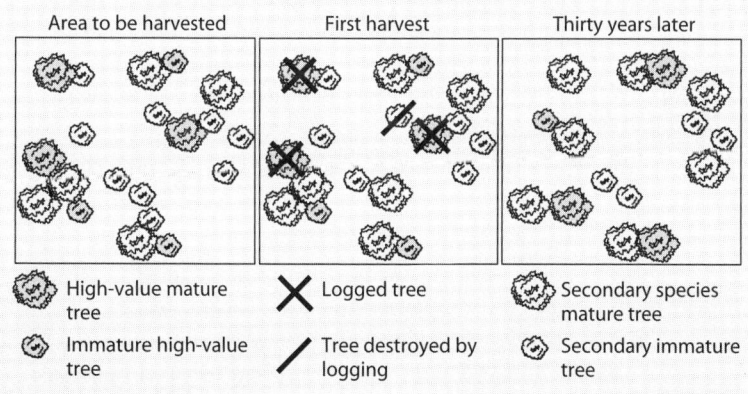

| Area to be harvested | First harvest | Thirty years later |

High-value mature tree

Logged tree

Secondary species mature tree

Immature high-value tree

Tree destroyed by logging

Secondary immature tree

Source: Authors.

Note: Fifty-nine percent of UFAs in southern Cameroon are estimated to have been harvested at least once already (Abt and others 2002).

harvests and improved marketing of secondary species offset the initial costs. Globally, financial returns on forest management can be attractive if efforts are made to expand the number of species harvested and to reduce costs through low-impact harvesting and the effective use of information from forest inventories.

Reforms in Forest Taxation

Until 1994, Cameroon's forest taxation regime was based mainly on taxes levied on exported logs; its chief objective was to capture revenue. The tax regime introduced by the reform responded to the equally important objectives of promoting sustainable forest management, increasing local processing, facilitating tax collection, sharing forest rents more equitably by widening the participation of communities, and improving governance and transparency in the forest sector. The new taxation system was also designed to at least partially offset the significant reduction in fiscal revenue expected from the ban on log exports that took effect in 2000. For a chronology of tax reform, see appendix 5.

Changing the Fiscal Regime to Improve Forest Management

To achieve the objectives that have just been described, the new fiscal structure (figure 3.5) relies mainly on:

- Shifting the tax basis from the product (volume of timber felled, processed, and exported) to the area of the concession (in the form of a competitively determined area tax to be paid annually regardless of harvested volume). This shift, which made industry pay a significant price for accessing forest resources, was expected to discourage speculation and produce predictable income streams for the state and for local communities that were easy to determine and collect.
- Introducing a sawmill entry tax to help control timber flows and discourage waste.
- Shifting the bulk of taxation away from exports and toward timber operations. This shift was expected to create incentives for forest management (with an increase in the number of harvested species and a reduction in timber waste), marketing innovation, and increased processing efficiency.

The changes in fiscal structure and the creation of the Enhanced Forest Revenue Program (Programme de Sécurisation des Recettes Forestières, PSRF) helped, along with the auctioning of harvesting rights, to maintain considerable fiscal revenues despite declining volumes of production and profound changes in the ratio of processed wood to logs (which changed from 32:68 in 1999 to 89:11 in 2006). Revenues from area taxes rose from $1.1 million (CFAF 260 million) in 1991–92 to $30.6 million (CFAF 15.3 billion) in 2004. In 2005, the three main forest taxes (area tax, felling tax, and sawmill entry tax) represented more than 88 percent of the total

Figure 3.5. Changes in the forestry tax structure

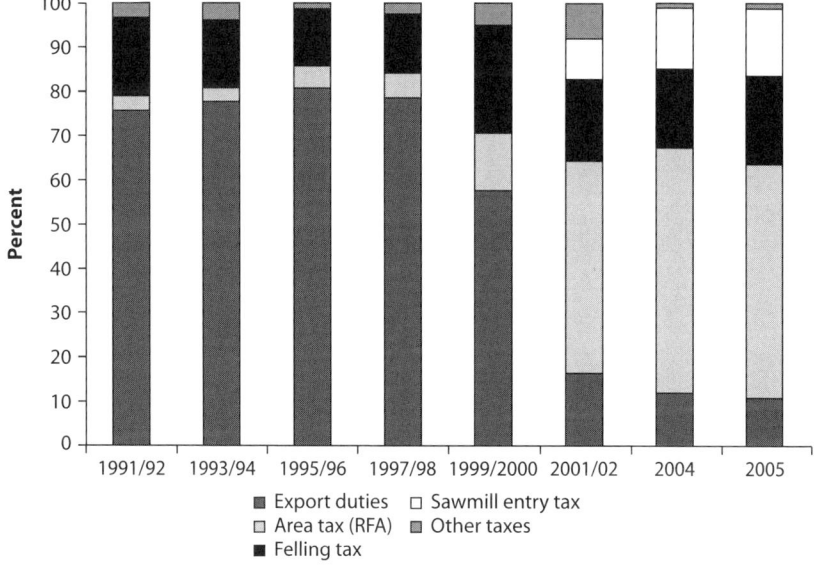

Source: Authors, based on data from PSRF and MINFOF.

forest tax burden, and the export tax represented about 10 percent. The decline in revenues between 1997/98[5] and 2005 (figure 3.6) appears modest, considering that over the same period commercial harvests declined by more than 32 percent, from 3.4 to 2.3 million cubic meters.

Guarantee System to Secure Fiscal and Environmental Performance
The 1994 Forest Law included a provision to secure compliance with fiscal and environmental commitments through a guarantee system. When awarded a concession, a firm would provide a guarantee bond equal to the amount of the annual area tax. If a company failed to meet its fiscal or environmental obligations, the Treasury would activate the guarantee and recover the equivalent of the unpaid sum. Despite the system's simplicity and potential to positively influence companies' behavior, the Treasury has not activated the guarantee even in cases of well-documented illegal logging. Most companies paid the area tax regularly, and the guarantee created a financial burden and reduced the ability to obtain credit. A few well-connected companies were able to delay posting bank guarantees until public opinion placed sufficient pressure on the Ministry of Finance. To sum up, it was a good idea, but it did not work as planned.

Figure 3.6. Forestry tax revenue, 1991–2005

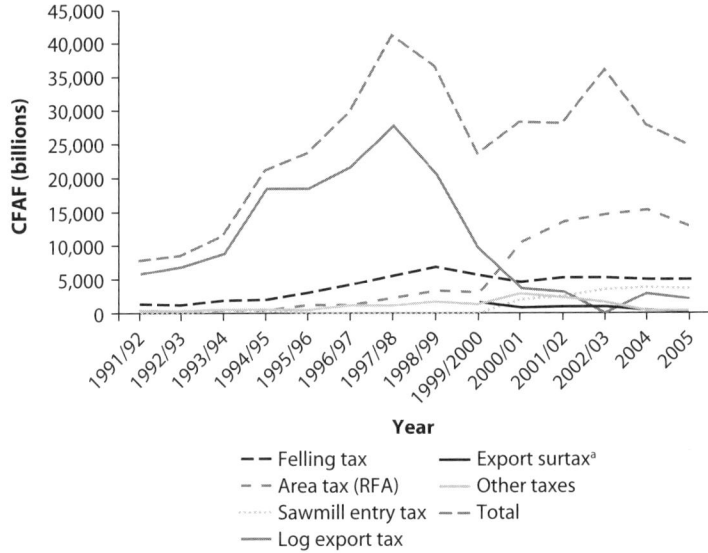

Source: Authors, based on data from PSRF and Customs.

a. Introduced in the 2000/01 Finance Law to replace the progressive surtax.

Distribution of Fiscal Revenues to Local Councils and Communities

The redistribution of forest revenues is a fundamental means of enabling local administrations and communities to benefit from forests and start to perceive them as a productive asset that is worth preserving. The July 1998 Finance Law mandated that 50 percent of revenues from the annual area tax must be distributed to local councils and communities; councils would receive 40 percent and communities 10 percent.

The area tax is recovered by the Enhanced Forest Revenue Program, a collaborative program of MINEF and the Ministry of Finance. Funds are transferred to the mayors of local-level governing bodies (councils) located in proximity of forest concessions based on local development plans that have ostensibly been drawn up with the respective communities. To increase the transparency of this process, the World Bank and its partners have strongly urged that the amounts provided to the local councils be published in the local and national press.

The Enhanced Forest Revenue Program redistributed about $66 million (CFAF 35.4 billion) to local councils (80 percent) and communities (20 percent) between mid-2000 and the end of 2005. These funds represent the primary source of income for communal councils. Major difficulties remain, as revealed in the course of two audits (a third was under way in 2008). Some of the problems are linked directly to administrative weaknesses, such as overlapping boundaries of local council and community territory (these boundaries are the basis for calculating the funds allocated) and inconsistencies in the databases of the Ministries of Forests and Finance.

Improvements can be introduced, but difficulties in establishing a fair and efficient system to allocate forest revenues to local governing bodies and communities are likely to persist in the context of limited accountability and poor local governance. Rural populations have little confidence in the state's impartiality and commitment to the collective interest, and they perceive government institutions as instruments for diverting public funds into private hands. The work of the recently created *Court des Comptes*, the independent public body in charge of auditing public institutions, may address these deeply rooted perceptions.

Institutional Reforms

Institutional reform was integral to wider reform in the forest sector. Some institutions had mandates that were inconsistent with the new policy that government should withdraw from production and marketing. In other institutions, corruption was endemic, and collaboration with other institutions and external partners was essential to gain credibility. Institutional reforms sought to:

- Focus institutions on their public service functions
- Align sector organization and sector expenditures with policy objectives
- Expand accountability beyond MINEF
- Create partnerships to monitor forests

Each of these objectives is discussed in the sections that follow.

Focusing on Public Service

In the early 1990s, several ministries and agencies were involved in managing Cameroon's forests. The Ministry of Agriculture administered commercial forestry, the Ministry of Tourism administered protected areas, and parastatals carried out a variety of productive and commercial functions.[6] Environmental issues were generally not included among the responsibilities of public or parastatal organizations and were generally not taken into account.

The 1994 Forest Law required that forest institutions withdraw from productive and commercial activities and enabled the Ministry of Forests to enter into collaborative agreements to obtain services provided by the private sector, communes, local communities, and NGOs. Reorganization of the National Office of Forest Development (Office National de Développement des Forêts, ONADEF) typified the broad need to align forest institutions and their functions with the new policies. ONADEF had been created before the first ministry in charge of forestry (which was MINEF, in 1992) was established and before adoption of the 1994 Forest Law. A parastatal in charge of industrial reforestation and promoting timber use, ONADEF had a monopoly on forest inventories and management plans, a status that ran counter to the development of consulting services in the private sector and fueled the conviction that corruption and clientelism dominated its workings. Resistance predictably delayed the process, but ONADEF had no place in the new institutional setting and was eventually dissolved in 2002. A review of forest institutions in 2002 strongly influenced preparation of the National Forest and Environment Sector Program (FESP), which helped to speed institutional change (including budgetary change; see the next section). ONADEF was succeeded by the Agence Nationale d'Appui au Développement Forestier (ANAFOR), whose mandate is to promote financially and environmentally viable plantations established, owned, and financed by individual farmers, communities, and the private sector.[7]

Aligning Sector Expenditures with Policy Objectives
Until the FESP began in 2004, national budget allocations to the Forest Administration were quite modest and did not reflect its changed functions. Funding for the core functions of planning, control, and law enforcement remained largely limited to wages. Any major innovations were funded either from extrabudgetary, ad hoc resources provided by the Ministry of Finance or donors. ONADEF—despite its clear irrelevance in the context of the reforms—absorbed half of forest sector funds until its demise (I&D 2002). The cash pumped into ONADEF's reforestation programs and manufacturing activities created immense opportunities to sustain the patronage system.

In 2004, with the implementation of the FESP and support from the international community, a major effort began to align the budget to forest policy for the first time (box 3.7). Budgetary planning directly reflects the objectives specified in the 2005 Forest and Environment Sector Policy Letter (appendix 6). A Medium-term Expenditure Framework has been adopted in which the headings and corresponding expenditures are clearly related to those in the policy letter. Although financial resources are now

Box 3.7

Cameroon's Forest and Environment Sector Program

Cameroon's Forest and Environment Sector Program (FESP) reflects the conviction that forest management, biodiversity conservation, social equity, and good governance are highly interdependent goals that must be approached simultaneously. The program reinforces the forest reform agenda by explicitly aligning expenditures with policy objectives for the forest and green environment.

Developed by the government of Cameroon, nongovernmental organizations (NGOs), the World Bank, and other development partners, the FESP is a 10-year program, adopted in June 2004. It provides a common sector development framework for government institutions, local governing bodies, communities, the private sector, NGOs, and partners active in the forest domain. Under this framework, donors support a consolidated development effort. Projects under way in the forest and environment sector have been merged or repositioned to fit FESP objectives and implementation structures.

The FESP's components, listed in the table below, deepen and scale up the reforms initiated with the 1994 Forest Law and the Third Structural Adjustment Credit for Cameroon. The program gives particular emphasis to strengthening national institutions and civil society in ways that permit local communities and the private sector to engage gainfully in the sustainable management, conservation, and development of forests and other natural resources.

Building capacity.   Perhaps the most important part of the program is to rejuvenate the human resources who manage Cameroon's green resources. Under FESP, the Ministry of Forests (MINFOF) and the Ministry of the Environment and Protection of Nature (MINEP) are recruiting 1,550 new staff: 600 in 2006, 400 in 2007, 300 in 2008, and 250 in 2009, all qualifications included. This process compensates for the projected retirement of two-thirds of forest and environment ministry staff, thereby improving staff quality and motivation without significantly increasing the salary bill. Staff recruitment is complemented by a training program as well as by a program to rehabilitate infrastructure, supply new equipment, and reconstruct management systems for personnel, assets, inventory, material flows, and financial resources.

Box continues on next page

FESP component	Subcomponent	Estimated cost ($ millions)
Environmental regulation and information management	Environmental regulations	7.5
	Environmental monitoring	
	Communications and environmental awareness	
Production forests management	Completion of the country zoning plan	30.7
	Forest management plans	
	Wood product industrialization	
	Control operations and sanctions	
Protected area, biodiversity, and wildlife management	Biodiversity planning and zoning	33.2
	Knowledge and information management	
	Participatory protected area and community wildlife area management	
	Design and implementation of protected-area management plans	
	Optimization of economic benefits of protected areas and hunting zones	
	Legal and institutional reform of protected-area and hunting-zone management	
	Sustainable financing of wildlife and protected areas	
	National Biodiversity Strategy and Action Plan update	
Community forest resources management	Capacity building for community forest and natural resource management	26.4
	Reforestation and forest regeneration	
	Promotion of fuelwood supply	
Institutional strengthening, training and research	Transition of ONADEF into ANAFOR	87.5
	Strengthening MINEP	
	Rehabilitating education and research in the forest and environment sector	
	Capacity building and decentralization	

Box 3.7—continued

Decentralizing and refocusing government institutions.   Through the FESP, MINEP and MINFOF are strengthening their oversight and regulatory functions at the center and in the field while relinquishing such functions as direct production, marketing, or other work that can be handled more effectively by the private sector, local communities, or NGOs. This shift began as the FESP was being prepared, when the National Office of Forest Development (ONADEF) became the National Forest Development Support Agency (ANAFOR). Plans were developed to recruit and train new staff to support ANAFOR's new functions. More than 500 forest workers whose jobs were no longer relevant were terminated.

Fostering greater transparency and disclosure.   The interrelated goals supported by the FESP require information to be gathered and shared readily and widely. The FESP supports information sharing in a number of ways. It defines regulations that meaningfully guide environmental monitoring and impact assessments.

It stresses the creation and dissemination of knowledge to inform the local use of resources (examples include multiple resource inventories and cartographic materials). Through its strong environmental awareness component, the FESP also helps people to value and use these knowledge resources.

Independent observers foster public confidence in and improve the public image of forest policy. Observers work with the government to control commercial logging, detect illegal logging, and ensure that the public auction of forest concessions proceeds correctly. The program supports the issuance of public reports detailing infractions and explaining how penalties were determined.

better aligned with policy objectives, it remains challenging to ensure that channels are developed and used to carry out policies on the ground.

Expanding Accountability beyond the Ministry of Forests

During the late 1980s and 1990s, the Forest Administration (which was part of the Ministry of Agriculture until MINEF was established in 1992) had come to control all forest-related information. It was largely perceived as a gatekeeper on behalf of its own interests and the interests of higher authorities and their clients. It soon became clear that this institution would be unable to reform from the inside. As the forest sector was increasingly drawn into the wider reform agenda, reform-minded government officials and the World Bank expanded the number of institutions at the table when forest-related decisions were made and widened their roles in activities that were once the exclusive domain of the Forest Administration. The Ministry of Finance has become a key actor in forest reform, for example, through the creation of the joint Enhanced Forest Revenue Program, which ensures that revenue is fairly and systematically collected and channeled to the Treasury, local governing bodies, and local communities.

The program has removed MINEF from a sensitive and lucrative area and brought new skills and perspectives to an essential public service. It greatly facilitates implementation of the tax reform and increases forest tax recovery rates. The program was restructured in 2001 to streamline and centralize payment of all forest-related taxes, value added taxes, and income taxes paid by forest companies. Because the Ministry of Finance is the dominant force in the program, however, the Ministry of Forestry is reluctant to cooperate, although it is the only source of accurate technical information for tax collection. As a consequence, the Ministry of Finance has sought to develop its own information system, using a method appropriate for corporate taxation (with a self-declaration system and random ex post controls) but inappropriate for forest taxation (especially because it does not link tax payments and verification of legality).

Creating Partnerships with Independent Observers to Monitor Law Enforcement

Supporters of reform within the government pushed to develop partnerships with internationally respected NGOs to perform two functions critical to the reforms' success. WRI agreed to monitor the status of forests by detecting illegal logging in forest concessions and protected areas through satellite image interpretation and information dissemination. To support enforcement of forest laws in the field, GW was asked to collaborate with MINFOF control teams. These partnerships strongly influenced the international credibility of forest sector governance in Cameroon and discouraged questionable behavior (box 3.8).

Independent observers are now a well-established institution in the forest sector; their presence on the Harvesting Rights Award Commission and in Forest Control Brigades is widely acknowledged. Independent observers have helped reformers impose order on systems that had become so corrupt that not even orders from the top echelons could be

Box 3.8

Partnerships with international nongovernmental organizations

Independent forest monitoring

In June 2002, the government of Cameroon signed a formal agreement with the World Resources Institute–Global Forest Watch to jointly monitor all timber harvesting nationwide. The monitoring, based on remote sensing and geographic information systems, includes such tools as interactive maps and databases consistent with original Forest Department data; field observations; and comparisons of data with the private sector and national nongovernmental organizations. The information gathered is verified with all stakeholders before being officially accepted by the Ministry of Forests and Wildlife (MINFOF) and released on the Internet as the annual *Interactive Forestry Atlas for Cameroon*. The atlas has become a powerful tool for governance, because it discloses detailed data on forest harvesting permits, maps the location of harvesting in all types of forests, and reveals the presence of forest roads and other physical signs of legal and illegal logging.

An independent observer to support law enforcement

Independent monitoring by third parties has been used in Cameroon since 2001 to detect and prosecute illegal logging more effectively. The establishment and maintenance of independent monitoring was a trigger for disbursement under the Country Assistance Strategy and for debt relief under the Heavily Indebted Poor Countries Initiative.

The monitoring project was not initiated (though it was later cofinanced) by the World Bank; rather, Global Witness (GW) was invited to carry out this groundbreaking work in 2001 after successful scoping missions funded by the United Kingdom. GW was appointed independent observer in May 2001 and served in that capacity until March 2005, when it was succeeded by Resource Extraction Monitoring (REM), an internationally recruited nongovernmental organization. Among other things, the independent observer is responsible for:

- Ensuring the objectivity and transparency of monitoring undertaken by MINFOF
- Strengthening the operational capacity of MINFOF's law enforcement services, especially the Central Control Unit (CCU), by improving and enforcing procedures

Box continues on next page

Box 3.8—continued

- Clarifying the roles of different players in forest monitoring and following up on offenses and sanctions, based on the legal and regulatory framework
- Helping to monitor the implementation of recommendations and decisions from CCU missions undertaken with the independent observer

Between June 2001 and March 2005, GW undertook 171 missions as independent observer. During its tenure, GW reported consistent violations of the law. Although infractions declined temporarily within UFAs, eventually out-of-boundary logging seemed to have been replaced by overharvesting certain species within concessions. From April 2005 to February 2007, REM visited more than 100 forestry operations (UFAs, ventes de coupe, community forests, and small titles). REM noted a reduction in logging in managed forests but a sharp increase in irregularities related to the origin of timber, such as false declarations of measurements or species in transport bills. More than 63 percent of infractions fell into this category.

executed. Relations between independent observers and the groups they monitor, especially the Forest Control Brigade, have been and are likely to remain tense, but they have yielded results and are expected to continue to do so. Their reports increase public interest and awareness, notably through a network of local organizations that widely disseminate the results. The government's willingness to submit to the highly public and often severe criticism of the independent observers signals an unwavering commitment to transparency.

Notes

1. At one time, a portion of the concessions was reserved for bids by Cameroonian nationals. This measure, lifted in 2001, was reinstated during the last bidding rounds.
2. The floor price is $4.20 (CFAF 2,500) per hectare per year for ventes de coupe and $1.70 (CFAF 1,000) per hectare per year for UFA concessions).
3. Although this outcome is clear, its causes are difficult to identify with precision. Bids may have risen because firms wished to secure

rights to certain concessions, wished to ensure sufficient supplies for their recently installed processing facilities, recognized that the process was more competitive, or recognized that the auction offered an opening for firms without ties to the old patronage system.

4. Note that these are not the same as the simple management plans developed for community and local council forests.

5. Timber production peaked in 1997/98 as a direct consequence of buoyant Asian demand, combined with the impact of the 1994 devaluation of the CFAF.

6. These parastatals included ONAREF (National Bureau for Forest Regeneration, Office National de Régénération des Forêts), CENADEFOR (National Center for Forestry Development, Centre National de Développement des Forêts), and SOFIBEL (Bélabo Lumber Products Corporation, Société Forestière et Industrielle de Bélabo).

7. There was considerable resistance to the establishment of ANAFOR, which lacked staff and operating funds until fairly recently but now plays a major part in a government plantation program.

PART II

Consequences of Reforms

CHAPTER 4

A Summary of Impacts

The overarching objective of Cameroon's forest sector reforms was to benefit great numbers of people and the environment by replacing chaotic and opaque arrangements for accessing forest resources with a more organized, transparent, and sustainable system. The previous chapter has given some indication of the reforms' scope and complexity. But what actually happened in the rainforests over the past 10 years? Although too many variables are involved to establish definitive causal relationships between individual reforms and specific changes in forest policy, and between the reforms in their entirety and overall transformation of the forest sector,[1] it is clear that some sector settings changed profoundly and that new trends emerged. This chapter very briefly summarizes these changes and assesses their combined effect. For the data and analyses behind this assessment, see chapter 5.

General Trends

Cameroon has moved toward meeting the reforms' objectives, although setbacks and unforeseen challenges are still taking place in a process that appears far from complete.

Some of the positive impacts of forest reform are immediately visible. Maps of Cameroon's forests between 1992 and 2007 clearly reveal changes in the dimensions and types of forests (map 2; see map insert). The incredible patchwork of short- and long-term, small and large, and often overlapping logging titles that prevailed in 1992 has been eliminated. New protected areas, community forests, and community hunting zones have been established and are expanding rapidly. Finally, deforestation remains modest overall in Cameroon, occurring mostly in the central regions and in the nonpermanent rural domain, where population is increasing rapidly and conversion of land to other uses is a legally acceptable option.

At the same time, as increased control, higher costs, and more discerning international markets helped reduce illegal logging by industry, companies started selling the totality of their production overseas. Legally sourced timber practically disappeared from the local markets. New informal producers gradually emerged, meeting the demand for timber from unauthorized sources and making illegal logging the normal way to supply timber for local markets.

Many newly created community forests were diverted from their original intent, served the interests of the few, and had to be suspended. The range of species used by industry did not expand as it was hoped, and this trend alone threatens to undermine the feasibility of sustainable forest management. Innovative governance and technical solutions such as the competitive award of concessions and stringent planning of harvesting rights had to be continually adapted to frustrate strategies for circumventing them. Cameroon's deliberate effort to use the forest sector as a pilot to raise the governance bar created the expectation that this sector would instantly become invulnerable to corruption and governance failures—an expectation that was impossible to fulfill but left the sector and its administrators more exposed to criticism.

Despite occasional setbacks and inconsistencies, much has been learned, and the reforms have created a framework for forest governance and sustainable management that is far more effective than the previous framework. There is evidence that inroads have been made in the corruption endemic to the forest sector (box 4.1). It is also clear that a significant unfinished agenda, ultimately with very high stakes, remains. This agenda is taken up in chapter 6.

Specific Observations

Following are specific observations on progress, success, and shortfalls.

Sector Transparency

The disclosure and dissemination of public information were the most dramatic changes in the forest sector between 1998 and 2008, and they have had dramatic results. The improved availability of public information has heightened the political and financial risk associated with the old influence and clientele system. With the involvement of external

Box 4.1

Cameroon, corruption, and the forest sector

In the Corruption Perception Index issued by Transparency International, which ranks countries according to general perceptions of corruption in the public sector, Cameroon ranked last among 85 countries in 1998 (last place in the last quintile). In 2007, Cameroon's ranking had changed to 138 out of 180 countries (lower end of the second-to-last quintile). This achievement is substantial, even though Cameroon has a long way to go to further reduce the perception of corruption.

Although the Corruption Perception Index does not reflect changes related to specific sectors of government, a recent Transparency International study of Cameroon identifies the forest sector as one of the country's "poles of corruption" (Transparency International 2007), observing that "The World Bank has done much to reform and clean up forestry policy in Cameroon. Its action in a way forms the cornerstone of subsequent governmental and civil society involvement" (p. 27).

Source: Transparency International 2007 and various years.

partners, forest ministry officials can no longer serve as the inscrutable gatekeepers of harvesting rights, allocating them to serve their own interests and those of allied institutions and elites.

Disclosure and dissemination of information have also introduced new elements of democracy and equity in the forest sector. Previously it was possible to circumvent competitive allocation procedures and award large forest concessions to influential individuals, but increased public interest and information made this and similar operations more difficult. Diverging private and public interests and a broader range of actors claiming a share of power and revenues created a new demand for governance in the forest sector. Decision makers gradually began to take into account the expectations of increasingly well-informed and active external and internal constituencies (especially civil society organizations).

Protected Areas

Very significant progress has been made in protecting Cameroon's forest biodiversity since the early 1990s, when the Campo Ma'an reserve was logged illegally, openly, and with impunity. Since 1995 Cameroon has expanded its network of protected areas, which occupies nearly

4 million hectares. It has also taken steps to develop and adopt management plans to ensure their protection. Between 2003 and 2008, Cameroon greatly increased the effectiveness with which it manages protected
areas, and their management is increasingly linked to the management
of public and private adjoining areas. The strong provisions for community involvement contained in recently developed management plans for
the protected areas of Campo Ma'an and of M'Bam and Djerem are
encouraging.

The financial sustainability of all of Cameroon's protected areas
remains weak, although efforts to integrate new conservation options
such as carbon offsets and conservation concessions may help.

Community Forests

Community forestry made its appearance and developed rapidly through
the reform period, but it remains very much a work in progress. Since
1998, when the first community forest manual was prepared with the U.K.
Department for International Development, more than 400 requests to
establish community forests have been made. Forests on 1.3 million hectares were set aside, but community interests were often undermined by
vested interests. Seventy-five community forests are now in operation.

Different models of community forestry have been adopted, such as
subcontracting industry to harvest community forests, restricting community forest use, and permitting artisanal logging based on long-term
harvesting cycles. Each of these models has limitations, and none has
worked better than the others. Burdensome, costly, and disempowering
regulations are among the most common causes of concern or failure.

While much remains to be done, observers concur that community
forestry in Cameroon is 10 years ahead of similar efforts in other parts
of Africa's humid tropics and that lessons from Cameroon will be useful
to other countries (Julve and others 2007). Once the object of residual
attention from institutions focused on industrial forestry, community forestry is now one of the most-funded forest subsectors.

Illegal Logging

Prior to the late 1990s, Cameroon exhibited neither the political will nor
the law enforcement capacity to curb illegal logging. Today the situation is
quite different. There is solid evidence that illegal logging declined significantly in absolute terms after 2002, most significantly in the permanent
domain, where forests are subject to management plans and adherence to
plans is gradually taking root. Illegal logging now largely supplies domes-

tic and regional markets (Sahelian and North African countries) rather than major export markets. The perpetrators are mostly small-scale operators who emerged after the devaluation, as industry gradually stopped supplying local markets and neglected local demand for timber.

Use of Other Forest Products

Although illegal logging draws a great deal of national and international attention, special products licenses, which permit the use of nontimber forest products such as fiber and traditional medicines, remain a major concern because of the lack of transparency with which they are awarded and poor monitoring. Allocations are based not on resource inventories but on requests from exploiters, heightening the risk of unsustainable use.

Sustainable Management of Production Forests

Forest management requirements have limited the previously blatant and unregulated overexploitation of Cameroon's forest resources, reduced the timber industry's environmental footprint, and placed forest management on a more sustainable basis. Between 1998 and 2007, the proportion of Cameroon's production forests covered by forest management plans (66 percent) became one of the highest among tropical countries. Forest management plans are now systematically prepared by all forest companies, reviewed by the administration, and made operational. Map 3 (see map insert) depicts the various uses of Cameroon's production forests and the status of management plans in 2007.

The size of the area open to logging each year was halved between 1998/99 and 2006, and the total harvested volume declined significantly. More timber is now harvested from UFAs, which are subject to more rigorous regulation, than from rural areas, where logging is less regulated and more destructive.

The adoption of management plans did not significantly enlarge the range of harvested species as expected. This failure threatens sustainable forest management and the sustainability of commercial operations. Without strong measures to promote the use of a broader range of species in a broader range of products, industry will continue to focus on species that compensate for higher production costs.

Third-party certification has progressed rapidly. Adoption and adherence to management plans has facilitated certification, which enables timber companies to reach an increasingly discerning and demanding market. Even more important, certification is helping government to overcome the limitations of traditional forest control by shifting the burden of checking for compliance with technical and legal standards to industry and away from government. It is possible that more than 1 million hectares will have been certified by the Forest Stewardship Council alone in Cameroon by the end of 2008.

Industry Structure

Cameroon's forest industry changed markedly after sustainable management rules and the auction system were introduced. A significant part of Cameroon's forest industry has restructured, adapting its business model to cost structures that include higher fixed costs, new investments for management plans, and increased social and environmental responsibility. New alliances and partnerships have emerged. Rather than focusing almost exclusively on logging, industry gives greater emphasis to processing and producing higher-value products. This diversified and less monolithic industry seems less intent on maintaining old privileges and more open to change. The diversification of capital, the arrival of new investors, the acquisition of more concessions by the same investors, the demise of the old timber barons, and the affirmation of national companies are all generally positive features.

Employment has also increased. Smaller and informal enterprises supplying local markets have contributed to vibrant job creation in rural areas—although much artisanal activity appears to contribute to untaxed and illegal logging.

Effects of Fiscal Reform on Industry

Changes in the structure of forest taxation created a new balance between the taxes levied on forest resources and taxes on exported timber. Taxes on exports fell; taxes on forest resources rose through the area tax and felling tax. Taxes as a percentage of company turnover increased threefold between 1992 and 1998; since then they have been relatively stable at around 19 percent on average. Fiscal reform changed the variety of products as industry moved away from exporting logs to processing them. The tax on unprocessed logs for export remains the highest per cubic meter of product, serving as a disincentive to export unprocessed logs and an incentive to produce processed timber products.

Poverty and Livelihoods

The reforms have fostered greater recognition of the rights of communities and indigenous peoples to use forest land, benefit from forest resources, and participate in forest decisions. These issues received scant policy attention before 1998. Progress has been significant but mixed, and an ambitious agenda remains.

The adoption of management plans in production forests and protected areas has generally consolidated the customary rights of forest and indigenous people, but efforts must be made to expand and further secure these rights. It is especially important to improve indigenous peoples' involvement and representation in local governing structures and to develop better mechanisms for them to benefit from the use of forest resources.

Through community forests, considerable progress has been made in formally linking communities to forest land over which they have rights under traditional law. Less progress is evident in linking these rights to secure flows of revenue. Excessive regulation, technical and economic constraints, insufficient knowledge of the social setting, and vested interests must be overcome if more communities are to establish, manage, and sustain forests on their own, to support their own priorities. Similar issues affect the development of local council forests.

Local markets and small-scale enterprises are the source of numerous jobs—precarious but essential—for the poor, yet little effort has been made to assess the potential impacts of reforms on local employment, supply chains for the local timber market, and revenues. The decision to suspend small titles between 1999 and 2006 successfully limited illegal logging by industrial groups but revealed obstacles to the legal operation of small-scale timber enterprises.

Impacts on poverty and livelihoods may improve as progress in decentralization and local governance continues. It is encouraging that forest communities now participate in government and industry processes that had long failed to engage them. Local people's issues are increasingly taken into account in forest-related policy decisions.

Note

1. For example, the analysis also considers how the forest sector was affected by significant economic changes occurring over the same period, such as the evolution of export markets for Cameroon's timber (especially in China and Western Europe) and the impacts of the 1994 devaluation of the CFAF.

CHAPTER 5

In-Depth Analysis of Impacts

Impacts on Sector Transparency and Public Participation

Disclosure and Dissemination of Public Information

The disclosure and dissemination of public information heightened the political and financial risk associated with the old influence and clientele system and introduced much-needed elements of democracy and equity into the forest sector. Civil society and the private sector began to demand better governance in the forest sector. These changes are related to reform measures that increased the disclosure and flow of public information, such as the involvement of third-party observers and better law enforcement.

The availability of forest information in the print, broadcast, and online media increased exponentially after 1998. Unpublished surveys by German Technical Cooperation (Deutsche Gesellschaft für Technische Zusammenarbeit, GTZ) between 2000 and 2004 found that more than 2,500 articles on forest and forest-related issues were published in the Cameroonian press. In 2003, MINFOF created a Web site (http://minfof.org/) providing public information on laws and regulations as well as on sensitive issues such as illegal logging cases, fines paid, cases in court, amounts of area tax redistributed to local councils, and lists of valid titles. The quarterly disclosure of valid titles and illegal logging cases was required under SAC III, and the government not only satisfied but exceeded this requirement. MINFOF placed on its own Web site and widely disseminated the first and second version of the *Interactive Forest Atlas* produced by GFW/WRI, with detailed information on the management and status of production forests, biodiversity reserves, and community forests. The government of Cameroon has also recently become a member of the Global Legal Information Network (GLIN), a public database of official texts of laws, regulations, judicial decisions, and other complementary legal sources. GLIN members provide full published texts of these documents to the database, which increases the transparency and visibility of their laws and regulations. Cameroon has indicated that the registration of texts in the database would start with forest-related laws and regulations.

Reports by Cameroon's independent observers (GW and REM) are available in English and French at http://www.globalwitness.org/pages/en/cameroon.html and http://www.observation-cameroun.info/.

The quantity and quality of disclosure gained a momentum that proved easier to reach than to maintain. Starting in 2006, information was released in a less timely and complete manner, and the MINFOF Web site, once the window on the forest sector, has often been hard to open.

Greater Public Participation in Making and Implementing Forest Policy

Decree 95-531, issued in 1995, made public participation and consultation a new tenet of forest policy making and implementation. The decree required the Forest Administration to seek the participation of local communities in decisions affecting forests (article 5) and defined the instances when public consultation was mandatory. Following the decree, local commissions (including community members, private sector representatives, customary authorities, and government and elected officials) were established in each province and systematically involved in forest gazetting and degazetting (articles 19 and 20) and in consultations related to the environmental impact of infrastructure and other projects affecting forests.

With increased scope for public participation, civil society organizations have started playing a more active role in framing and monitoring forest policies and laws. For example, civil society participated effectively in reviews and public discussions related to the environmental impacts of the Chad–Cameroon Pipeline Project and other projects, including the proposed Lom Pangar Dam, which will affect the Deng Deng forest and wildlife reserve. Organizations such as the Cameroonian Foundation for Concerted Action and Training on the Environment (FOCARFE) have helped communities affected by the pipeline's construction to promote agroforestry and preserve biodiversity. The Center for Environment and Development (CED) became involved with monitoring social and environmental developments after the pipeline project was completed and has helped indigenous groups defend their traditional rights to the forest. Although consultations with local groups do not always feature the kind of participation envisioned in the legal framework—and in some cases the legal niceties are disregarded altogether—coercive or cursory participation does decrease when local NGOs are involved. Local NGOs have become considerably stronger in recent years and are becoming more active advocates for full local participation in forest decisions.

Impacts on Illegal Logging

Trends in Illegal Logging

Illegal logging has clearly declined since 2002, but in any assessment of illegal logging it is important to consider that "illegal" is a relative concept. What is illegal in one country may be perfectly acceptable in another, and practices once regarded as legal in a given country may become illegal as rules change or become more restrictive. A country such as Cameroon, with more stringent regulations and relatively better control, is more likely to reveal illegal logging than a country where rules and controls are more loosely applied.

The types of illegal logging and their impacts—environmental, social, economic—can also vary widely: The ecological impact of harvesting trees from a biodiversity reserve is not the same as that of a firm cutting timber outside its specified annual harvested area but within its concession. The larger point is that while numbers are important, they do not tell the full story of illegal logging, which must be carefully characterized and understood to be curbed. Estimates of the percentage of illegal logging in Cameroon are very sensitive to assumptions about the level of domestic consumption, and reliable data on domestic consumption and the informal sector are not available.

The formal sector.   In a comprehensive description of illegal logging in Cameroon by specific activity between 1990 and 2004, Cerutti and Tacconi (2006) found that illegal logging by *industry* (that is, for export) appears to have been significant between 1998 and 2002 but that it occurred at a level "well below" the 50 percent of the total harvest commonly cited.[1] They distinguished two periods in which illegal industrial logging had very distinct characteristics:

- The years from 1998 to 2002 saw illegal activity develop on a massive scale (figure 5.1); it could have represented *more than one-third of annual production*. Many forces contributed to the proliferation of illegal logging by industry at this time. The effects of the devaluation persisted; licenses had expired; small harvesting titles had been suspended; and new UFAs had not yet been allocated, because the World Bank awaited improvements in the very corrupt allocation system. Illegal logging occurred mainly outside the boundaries of UFAs or outside the annual harvestable area within a given UFA.[2]

• Since 2002, this situation appears to have improved. Harvesting rights have been allocated within nearly all of the productive permanent forest domain (more than 6 million hectares). Several concessions changed ownership and some processing units closed or changed hands, thereby eliminating most of the excess processing capacity that may have encouraged illegal logging. Some operators seeking international certification in forest management are regulating their activities more carefully. Others, coming under the watchful eyes of the independent observer and the Central Control Unit, have been cited for an increasing number of infractions. Data on declared industrial harvests and registered exports from the port of Douala tend to correspond (see figure 5.1), suggesting that illegal logging has diminished significantly compared with the previous period.

Official statistics may not tell the entire story, however. According to press accounts in France, a May 2008 report by the French organization Amis de la Terre (Friends of Earth) contended that illegal logging was widely practiced on small titles, which are granted at the discretion of Forest Administration authorities. This view echoes findings in the third annual report of the independent observer (REM) in April 2008, which

Figure 5.1. National timber harvest, exports, and domestic consumption

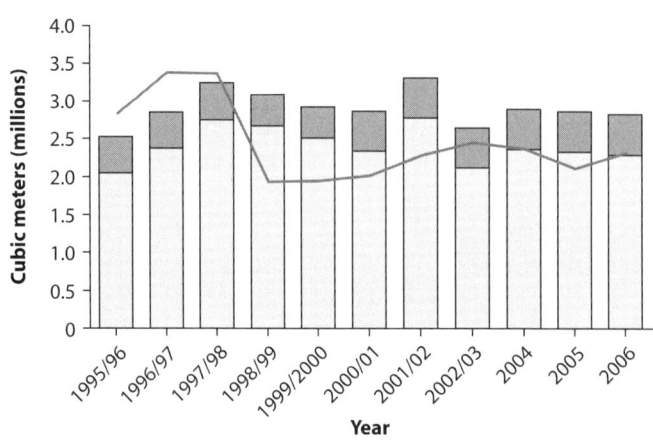

□ Exports (m³)   ■ Assumed domestic demand   —— Declared harvest (m³)

Source: Authors, based on data in Cerutti and Tacconi 2006 and updated by authors.

describes small titles as being "at the core of governance problems within the ministry of Forests."[3] According to REM, a subset of small titles called *autorisations de récupération de bois* (timber salvage authorizations) produced about 300,000 cubic meters of timber in 2006. Because the illegal provenance of most of this timber is hidden through falsified documents and routine administrative procedures, it cannot be detected through aggregate statistics such as those in figure 5.1.

Aside from being distorted by corrupt practices and vested interests, the data themselves are problematic. Cameroon lacks quality controls that would ensure consistency in the data collected. This deficiency simply compounds negative impressions of the integrity of information from the government. An encouraging sign mentioned in the REM annual report is that the head of MINFOF is aware of the issue and has undertaken several initiatives in recent months to address it, including staff reorganization and sanctions.

The informal sector.   As noted, logging traditionally was undertaken by industry based on a variety of permits, for small and large areas and short and long periods. After the 1985 economic crisis, many urban dwellers could not afford wood supplied by the formal sector. The situation worsened after the 1994 devaluation, which made timber exports twice as profitable (in CFAF) and caused industry to divert almost all of its attention away from a domestic market in which impoverished consumers could no longer afford legal industrial timber.

Until this time, artisanal loggers were so few and operated on such a small scale that their effect on markets was hardly perceptible, and little thought was given to regulating their activities. As industry departed from local markets and demand soared, artisanal loggers came into their own as a new actor in the forest sector. Their small size and local connections made control difficult. The fact that they were responding to a legitimate and irrepressible demand favored tolerance, which allowed them to expand rapidly.

Many forested areas outside concessions and community forests have been subject to illegal logging. Chain-saw operators are said to be numerous, although their number is unknown. Since the 1990s, Cameroon has imported a significant number (as many as 150) of mobile Lucas Mill saws, capable of processing 18,000 cubic meters per year.

Operating largely in the nonpermanent domain, the informal sector remains mostly illegal, and no coherent policy has been adopted to help small-scale loggers obtain timber from legal sources. Even though small titles are being issued again, many small-scale enterprises simply choose to avoid the bureaucratic delays and corruption involved in legalizing their operations.

Cerutti and Tacconi (2006) estimated that the informal sector harvests 540,000 cubic meters per year, whereas Plouvier and others (2002) suggested a figure of about 1 million cubic meters. If one assumes that the informal sector harvests 800,000 cubic meters and adds the Forest Administration 2004 production figure of 2.2 million cubic meters, a total of 3 million cubic meters of timber could have been produced in Cameroon in one year, of which as much as 25–30 percent may have been illegal, principally issuing from the informal sector and supplying the domestic market. Such figures must be viewed with caution, given the potential for overlapping and otherwise unreliable data, yet the general trends are very clear.

Much illegal timber (high-value woods such as bété, bibolo, bubinga, sapelli, and zingana) is used to make furniture. Furniture-making is a vibrant segment of the informal sector. The biggest market is in Yaoundé. Local construction is also a significant consumer of illegally harvested timber processed into planks.

Although illegal logging tends to be concentrated in the informal sector and to be directed mostly at the growing domestic market, some potentially illegal timber is entering the international market.[4] In 2005, more than 100,000 cubic meters of timber, sourced from "operators without concessions," were exported through Douala, but it is difficult to determine how much came from companies operating more or less legally on small titles and how much came from informal chain-saw operators. About 200,000–260,000 cubic meters of log-equivalent timber were channeled to Sahelian and North African markets through very active networks in northern Cameroon, which source timber from densely forested areas of Center and East Provinces (Koffi Yeboa 2005) (see map 4; map insert), although the proportion of illegally sourced timber in this northern export flow remains largely unknown.[5]

Prosecution of Illegal Logging

The credibility of a law enforcement system depends on its capacity not only to detect fraud and crime but also to sufficiently fine or punish offenders. Since 2000, the Ministries of Forests and Finance have intensified the identification and prosecution of illegal logging operations, with support from the independent observer on forest law enforcement. As mentioned, MINFOF issues a quarterly list of illegal logging cases,[6] fines paid, and cases in court. Figure 5.2 depicts the fines paid between 2001 and 2007. Data do not, however, show the payment of damages, so they appear low (new data are being provided by MINFOF and the Ministry of Finance).

Several sources of concern remain. The fines whose payment has been ascertained, CFAF 2,209,364,548, or about $4.4 million, are not insignifi-

Figure 5.2. Fines paid for illegal forest activities, Cameroon, 2001–07

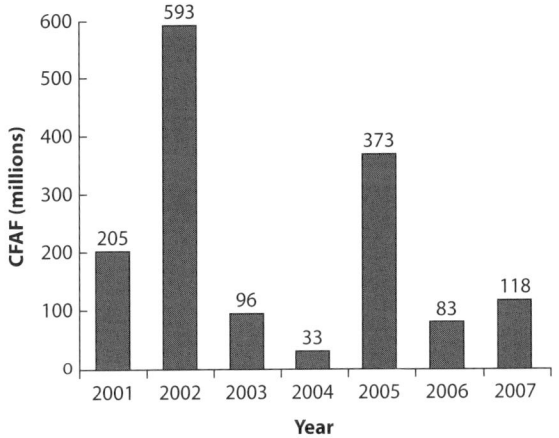

Source: REM, personal communication, 2007.

cant, but they represent a relatively small part of the fines issued between 2001 and 2007.[7] Some infractions cannot be prosecuted because companies have been dissolved and no assets remain. The Ministry of Forests still requires collaboration from the Ministries of Finance and Justice to prosecute cases diligently. To this end a joint committee was formed in 2004, but it never functioned.

In 2005, the independent observer called attention to the increasing number of forest cases that were resolved by mutual agreement between the Forest Ministry and individual companies.[8] (The regulations permit MINFOF to settle for reduced fines and damages whenever companies accept responsibility and are ready to pay without going to court.) The growing number of out-of-court settlements has become more worrisome because MINFOF has appeared to settle for amounts lower than the value of the timber illegally obtained by the companies, effectively undermining law enforcement and the prosecution of illegal activities. This practice attracted so much public attention that in 2008 MINFOF raised the level of such settlements to not less than 50 percent. According to MINFOF, a regulation mandating a minimum settlement is being prepared.

Impacts on Deforestation and Forest Conservation

Deforestation Trends

Recent and authoritative studies by the European Commission's Joint Research Centre, the Université Catholique de Louvain, and the University of South Dakota (de Wasseige and others 2009) show that average deforestation between 1990 and 2005 remained modest, around 0.14 percent per year. Loss of forest was confined to the central regions outside permanent production forests and national parks, occurring mostly in nonpermanent forest areas where population is increasing rapidly and conversion of land to other uses is a legally acceptable option. More refined sensing technology also allowed reassessment to 0.14 percent per year, down from 0.28 percent, of the gross deforestation rate previously reported for the 1990–2000 period.[9]

Progress in Protected Areas

Since the late 1990s, when forest companies logged the Campo Ma'an reserve with impunity, very significant progress has been made in establishing and protecting forest reserves. Protected areas have greatly expanded in size (map 5; see map insert). The effectiveness of management in the parks of Bakossi Kupe, Benoué, Boumba and Nki, Campo Ma'an, Korup, M'Bam and Djerem, Ndongoré, and Waza, measured with the WWF toolkit,[10] grew very significantly between 2003 and 2007.

The study documenting this improvement noted progress in integrating protected areas into the overall management of adjoining public and private landscapes. It cautioned, however, that indigenous people and local communities are not well integrated in the management of protected areas and that weaknesses remain in environmental education and the use of research in attaining park and landscape objectives.

Impacts on Integration of Global Environmental Services

As forest reforms gained ground, Cameroon started developing a more holistic view of its forests and their potential to offer global environmental services.

Although it included no meaningful provisions to support the conservation of natural forests, the 1997 Kyoto Protocol to the United Nations Framework Convention on Climate Change delivered the message that the value of forests far exceeded the value of the biodiversity they harbored and the building and manufacturing material they produced. New markets seemed to be in the making for standing forests to supply global environmental services. Only four years after Kyoto, Cameroon—betting on these emerging markets—started diversifying the options for its forest production estate by setting aside 13 percent of this area for potential conservation concessions (box 5.1). Located in the Ngoïla–Mintom area, these UFAs include some of the nation's best-conserved natural forests.

Box 5.1

Conservation concessions

Conservation concessions have been established in a growing number of countries, most notably in Latin America and the Caribbean. For example, Conservation International concluded an agreement in 2002 with the government of Guyana to protect about 80,000 hectares of forest for 30 years. In their simplest form, conservation concessions are agreements between national authorities or other owners of a particular resource to protect a critical habitat, especially of threatened plant, animal, or marine species, in exchange for compensation over time from conservation organizations or other investors. The proceeds can be used to achieve a number of goals—for example, to conserve an area effectively, to make conservation economically attractive for the local communities involved, to preserve options and maintain ecosystems until a formal system of protected areas can be established and maintained, and to compensate for alternative land uses that are inconsistent with conservation goals. Ideally, conservation concessions stimulate the long-term protection of natural resources and foster economic development at the same time.

One advantage of conservation concessions is that they provide a market mechanism for conservation. Because they resemble a standard business arrangement, conservation concessions can remove some of the political hurdles that often cause traditional conservation efforts to languish. The approach is also attractive because the results can be monitored based on verifiable norms, so investors can see what they are getting for their money. The norms usually spell out the balance between conservation and development goals, which may range from no development to traditional uses or to sustainable harvests of specific resources. Guidelines define the means for achieving this level of protection for the area, including standards for regulatory oversight and contingency plans to counter unexpected pressures on the area. Income from conservation concessions, unlike income from logging or tourism, is usually stable—which is especially good if the income is also being used to support economic development objectives. Finally, fair compensation of stakeholders is obligatory. Compensation cannot be limited to paying the government for forgone taxes. Jobs that are lost or never created because of the concession must also be taken into account, alongside the other local economic benefits that are generally associated with the timber industry, which can be expensive.

Conservation concessions are not a replacement for parks and other nationally protected areas. They are not ideal in every setting—for example, when it is vital to preserve an area in perpetuity, when payments are unlikely to bring the desired results, or when long-term funding is simply unavailable.

The Ngoïla–Mintom rainforest consists of nine UFAs extending across 870,000 hectares of unexploited forest—a vast area earmarked as production forest in the national zoning plan in the early 1990s. Conserving this forest would be a master stroke, because it would create a large international conservation corridor through national parks in Cameroon (Dja), Gabon (Minkébé), and Congo (Odzala) (see map 6; map insert).

The government's decision to refrain—at least temporarily—from auctioning harvesting rights to this forest was announced, enshrined in regulations, and reflected in all maps and other forest information. Finding a partner to sign the concession agreement has proven elusive, however. Are developed nations and eco-investors prepared to pay to protect forests that might be critical to the global environment? So far Cameroon has received only a few offers and virtually no credible ones. Early in 2008, it gave public notice (in the February 14 issue of the *Economist*) that it continues to be interested in conserving the area rather than turning it into a timber concession. Yet the government cannot indefinitely forgo the income that forest industry would secure and leave this area unmanaged and exposed to illegal logging, poaching, mining, and agricultural encroachment. The government and the WWF are working to devise a new, potentially more attractive conservation deal for Ngoïla–Mintom, including sustainable hunting and forestry on the fringes of the area and preservation of the core as a wildlife sanctuary. They are waiting for feedback.

The opportunity cost for establishing conservation concessions rather than timber concessions in this forest—which is enormous compared to most conservation concessions—is likely to be high. Karsenty (2007) estimated the net cost to be up to $14–15 per hectare per year, reflecting the fiscal revenue sacrificed by leaving the forest out of production and including compensation for lost employment opportunities. NGOs cannot pay these prices. According to the *Economist*, conservation NGOs would have considered even $2 per hectare, the amount supposedly mentioned by a high-ranking ministry official, to be too high.

Could Ngoïla-Mintom be conserved by renewed interest in making carbon markets work for African countries? The possibility of incorporating a mechanism to reward the reduction of emissions from deforestation and degradation (REDD) into the post-Kyoto architecture was raised at the 2007 United Nations Climate Change Conference. A system that rewards the reduction or avoidance of forest degradation could provide crucial incentives for conservation: Given that the Ngoïla-Mintom UFAs were originally earmarked for logging, their conservation would unambiguously represent an "additional" environmental benefit under a REDD mechanism (still to be defined) with the preservation of carbon stocks. Selective harvesting of these UFAs under a management plan would have yielded carbon emissions ranging from 40 to 120 tons of carbon dioxide per hectare harvested, depending on the intensity and method of logging. A conservation project in these same areas could be subsidized from carbon credits corresponding to the emissions avoided by not harvesting the timber as originally intended. Depending on the accounting rules (which remain to be decided but could include temporary credits, other assets, or possible discounting rules), the subsidy could be more or less large, and it could cover at least a part of the cost of the annual compensation to stakeholders. It would be the responsibility of the government and its development partners to ensure that all stakeholders are fairly compensated based on sound social and economic studies and on well-informed negotiation.

Impacts on Forest Management, Industry Structure, and Revenue

Trends in Sustainable Forest Management

Cameroon has become the most advanced country among its neighbors in placing production forests under approved management plans (table 5.1). The annual area open to logging fell by 50 percent between 1998 and 2006, along with the volume of timber harvested, although the limited range of species harvested indicates potential problems with sustainable management. An increasing number of companies are seeking third-party certification of adherence to rigorous management standards.

By early 2006, management plans for 55 UFAs (of the 72 allocated to 2002)[11] had been approved and were eligible for a contract upon completion of gazetting. These 55 UFAs cover 4 million hectares, more than 71 percent of the UFA area for which harvesting rights had been

Table 5.1. Adoption of sustainable forest management in Cameroon and neighboring countries, 2006 (thousands of hectares)

	Allocated to concessions	With an approved management plan	Percent
Cameroon	6,005	3,990	66.4
Central African Republic	2,920	650	22.3
Congo, Dem. Rep. of	15,500	1,080	7.0
Congo, Rep. of	8,440	1,300	15.4
Gabon	6,923	2,310	33.4

Source: Data for Cameroon from MINFOF. Data for neighboring countries from IITO.

Figure 5.3. Approval of forest management plans, 2003–06

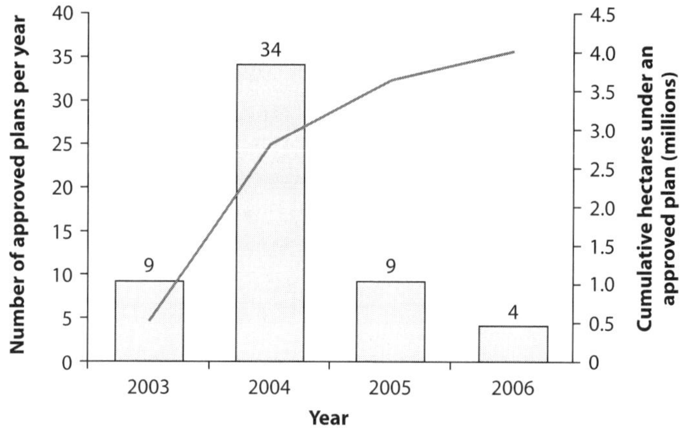

Source: MINFOF, April 2006.

allocated by 2005 (some 5.6 million hectares). Delays in approval before 2004 (figure 5.3) were occasioned partly by delays in gazetting.

Assessing the sustainability of forest management plans is challenging, even if they are well implemented (box 5.2). Assessments in Cameroon have focused exclusively on timber. The sustainable management of nontimber products, such as medicinal plants, hunting, fishing, or environmental services, has not been studied.

Box 5.2

What makes a forest management plan "sustainable"?

Can 30-year harvesting cycles such as those practiced in Cameroon be considered sustainable if the number of high-value trees never returns to its original level? It would be excessive to conclude in such a situation that management plans do not promote sustainable forestry. The forest transition—in which a first harvest including large numbers of long-maturing, high-value trees is succeeded by a harvesting cycle with fewer high-value trees and subsequent stable harvesting cycles—is a well-known phenomenon. It is routinely taken into account in forest management strategies and is not unique to the management plans adopted in Cameroon. Additionally, some species may become increasingly rare at the local level without being threatened more widely. Finally, industry can adapt to the changing species composition in several ways: by pursuing technical change (for example, installing processing facilities that use a wider range of species more efficiently), by fostering market changes that increase the value of "secondary" species (Karsenty and Gourlet-Fleury 2006), and by adapting the silvicultural strategy over time

What about carbon stocks in a managed forest? From a carbon standpoint, experimental results in the Congo Basin tend to indicate that aboveground carbon stocks recover within 25 years after a high-intensity cut (for Africa) of four trees per hectare and damage associated with extraction. Although the higher harvest intensity leads to greater carbon dioxide emissions during logging, it boosts the recovery of biomass and reduces the time necessary for full aboveground recovery. Carbon stocks are not the same as stocks of commercially valuable timber.

What about other aspects of sustainability? Different groups of stakeholders define sustainability in different ways, and their definitions have particular social as well as economic and ecological dimensions. Social sustainability—the effects of long-term forest management on local communities as well as forest industry employees—can be a concern. For example, a 2006 study by GTZ of a sample of 20 management plans showed that only half complied with the minimum legal requirements and that most showed weaknesses with respect to social and biodiversity issues (GTZ 2006). The report noted that government oversight on the quality of management plans remains wanting. The increasing number of firms that meet the requirement of the most demanding forest management certification schemes, however, suggests that the quality of management is likely to improve despite government tolerance of substandard plans. Greenpeace (2007) has contended that the management plans simply conceal widespread deforestation, although REM found that illegal logging has been reduced in managed production forests.

Does third-party certification guarantee sustainability? To gain third-party certification, companies are required to meet detailed social, environmental, and economic criteria, but certification does not resolve the debate over what exactly constitutes sustainable forest management. For example, Cameroon requires that forest management plans ensure that the most commercially exploited species recover to a minimum of 50 percent of their former level after 30 years. Certifiers sometimes require companies to attain a higher recovery rate (85 percent in the case of one company), although no serious scientific grounds exist to justify the alternative thresholds. Yet as rigorous as these certification schemes tend to be, they do not guarantee sustainability. The Forest Stewardship Council, for example, does not pretend that its *Principles and Criteria* (FSC 2002) defines "sustainable forest management" and carefully refers to "stewardship" rather than "sustainability."

Source: Authors; Fredericksen and Putz 2003; Karsenty and Gourlet-Fleury 2006.

Areas harvested.   As mentioned, new regulations limited the area that could be harvested to 1/30th of concession area each year. Total area where timber could be legally harvested has declined from more than 400,000 hectares in 1998/99 to 200,000 hectares in 2006. The reduction in area has been accompanied by an equally important shift in the type of management models used in the areas where timber was harvested, from mostly rural areas, where logging is less regulated and more destructive, to UFAs, where regulations are applied more rigorously (figure 5.4).

Volumes harvested.   The volume of legally harvested timber peaked in 1997/98 at 3.5 million cubic meters (figure 5.5). This level of production was a direct consequence of buoyant Asian demand, combined with the impact of the 1994 devaluation of the CFAF. Since then, harvested volume has come down and remained steady at about 2.2–2.5 million cubic meters, about 75 percent of which comes from sustainable sources (UFAs).[12]

Figure 5.6 highlights the origin of harvested timber: While most timber was extracted from unregulated areas in 1998/99, since 2004, more than three-quarters of declared production has come from concessions. In 2006,

Figure 5.4. Trends in areas open for harvesting, by type of harvesting right

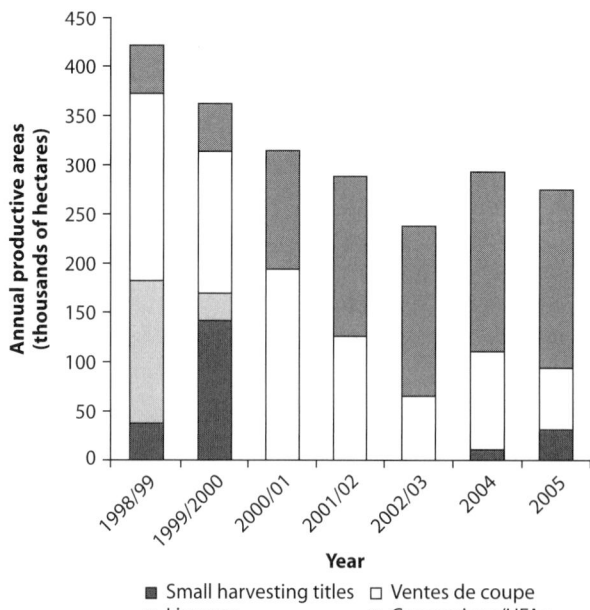

Source: SIGIF (MINFOF) unpublished data

Figure 5.5. Annual declared timber production, 1997/98–2006

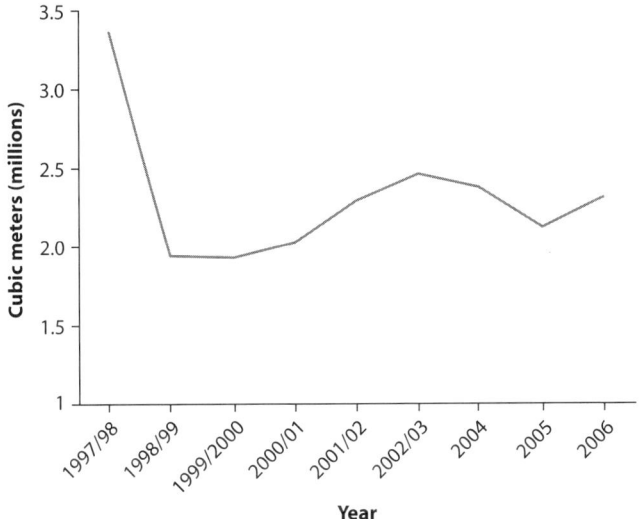

Source: Authors, based on SIGIF data (MINFOF).

the allocation of small titles was started again,[13] but measures were introduced to make allocation more transparent and permit monitoring (titles are no longer based on volume only but are assigned to clearly demarcated areas so that they can less easily become entry points for illegal logging). The decision to reopen small titles was a first response to local demand and a first attempt to bring small-scale illegal loggers into the formal sector.

Harvesting intensity and range of tree species harvested. Contrary to expectations, the harvesting intensity and range of harvested species did not increase. When average volumes are taken into account, harvesting intensities in UFAs and ventes de coupe do not appear significantly different in 1998/99 (10.8 cubic meters per hectare in UFAs and 4.2 cubic meters per hectare in ventes de coupe) and 2003 (10.0 cubic meters per hectare and 4.8 cubic meters per hectare) (appendix 7). This phenomenon probably results from a combination of factors, such as new limits on the size of harvestable trees and declining stocks of traditional commercial species in forests that often have already been harvested at least once.[14] It may also be an appropriate short-term response to a change in relative prices. Higher taxes and rising transport costs have brought the costs of less valuable species closer to free-on-board (fob) prices, reducing their margin of profitability compared with more valuable species.

Figure 5.6. Production by type of title and logging season

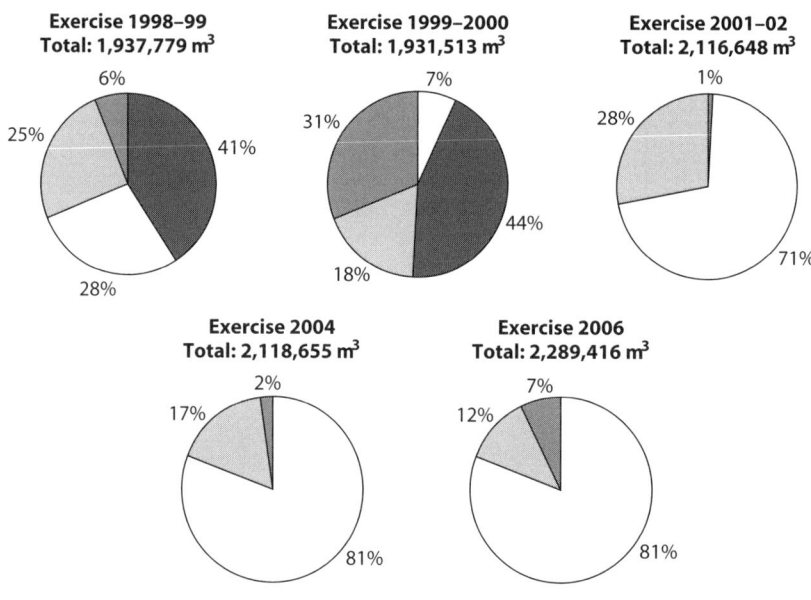

■ Licenses □ Concessions ▣ Ventes de coupes ▤ Special authorizations (AEB, ARB)[a]

Source: Authors, based on data from MINFOF and Customs.

a. AEB = Autorisation d'enlèvement de bois; ARB = Autorisation de récupération de bois.

In the absence of strong measures to promote secondary species (box 5.3), the reductions in average profitability will continue to offer incentives for companies to focus on species that can compensate for higher production costs. Unless industry can successfully market a greater range of species, the prospects for sustainable forest management will diminish.

Despite significant changes in total production in 1990/91, 1995/96, and 2002/03, the range of species used did not change significantly. Harvests remained very selective; two types of wood (ayous and sapelli) account for about 65 percent of the 10 most harvested woods (figure 5.7). Harvests of tali, okan (a substitute for azobé), and kossipo increased slightly, but the share of these less commercial woods remained marginal. It may be that fewer new species are harvested because Cameroon's forest industry has no further room to improve its technical or financial efficiency by modifying the species mix, yet there are many indications that this is probably not the case. It would probably be worthwhile to encourage the substitution of traditional commercial species with lesser-known species through fiscal incentives.

Box 5.3

No more primary forests available for logging

Fifty-nine percent of the UFAs in southern Cameroon are estimated to have been harvested at least once (Abt and others 2002). With practically all Cameroon's remaining primary forests protected from commercial activity, what are the implications for the timber industry and sustainable forest management?

Maps of the road network in Cameroon's forests indicate that it probably reaches between 1.2 million and 1.3 million hectares in southern and central Cameroon (GFW 2005) and GFW, personal communication). These roads make it very likely that over a large share of Cameroon's rainforest zone, timber will be harvested from secondary forests from which most high-volume trees of valuable species have been removed.

Harvestable volume has been observed to drop progressively in secondary forests (Vincent and Binkley 1992). The magnitude of this drop depends on the selectivity of the initial harvest(s), the diversity of the forest, and the industry's capacity to increase the range of harvested woods—which in turn depends on companies' ability to market lesser-known woods and to produce attractive products from them.

In Central Africa, where the initial harvests are highly selective, the drop in volume is less pronounced than in Southeast Asia, where the tree population is more homogenous from a commercial standpoint. If species such as frake (*Terminalia superba*) and tali (*Erythrophleum suaveolens*) are given greater consideration by industry, the volumes harvested in a second felling cycle can remain similar or even increase. If management plans are strictly respected, this shift to new species can sustain the forest industry and prevent the kind of brutal downsizing that has occurred in countries such as Indonesia.

Moving toward certification. Since 2004, as increasing numbers of management plans have been approved and implemented, the reports of independent observers indicate that compliance with basic forest management regulations has become the norm. This respect for Cameroon's demanding Forest Law and regulations has improved the prospects for forest companies to obtain third-party certification.

Certification enables companies to distinguish themselves in the marketplace as trustworthy partners of forest owners (governments) and buyers. Certification is purely voluntary and commonly assumed to make sense for companies only when there is a discerning market (in this case, a market in which consumers are willing to pay more for timber that is

Figure 5.7. Trends in tree woods harvested, various years, 1990/91–2004

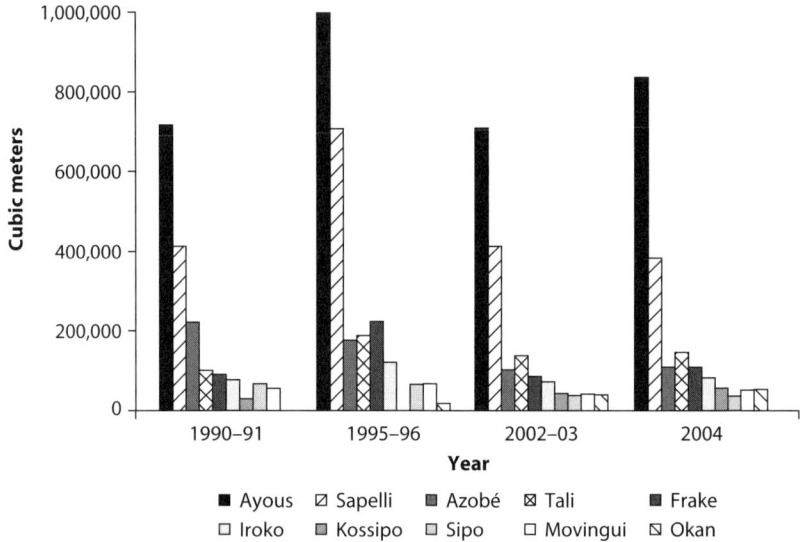

Source: Authors, based on unpublished data from MINEF and SIGIF.

produced sustainably and harvested legally). Although such markets have started emerging, certification is useful as more than a marketing instrument. In countries such as Cameroon, certification may play a useful public policy function by compensating for weak law enforcement and shifting the burden of control, at least partly, to certification bodies and to the companies themselves.

Companies accounting for more than 70 percent of Cameroon's forest exports have obtained certification of the legal origin of their products from EuroCertifor. A UFA of about 42,000 hectares, held by Wijma, was the first natural forest concession in Africa certified by the Forest Stewardship Council, the world's most demanding forest certification body. Many other applications have been submitted to FSC, the International Organization for Standardization (ISO), and other recognized organizations. In 2008, as much as 900,000 hectares were certified by FSC. This situation is a vast improvement over the uncertainty about the legal provenance and environmental footprint of timber exported by Cameroon just 10 years ago.

Impacts on the Forest Industry

Changes in industry structure.   Industry structure and product mix have changed significantly over the course of the reforms. In 1990, the forest industry predominantly focused on logging.[15] To comply with the wood-processing quota enforced between 1994 and 1999, and in response to the log export ban in 1999, practically all companies started to build processing units and integrate logging and processing. Timber-processing companies, hoping that the log export ban would make it possible to purchase logs from the local market, were disappointed when most companies processed all of the timber they acquired. Beginning in 2000, processors started to seek UFA concessions and ventes de coupe to secure their wood supply.

The modest size of Cameroon's UFAs (averaging 67,000 hectares) increases the fixed costs of forest management and restricts marketing flexibility. To offset the risk inherent in the small size of UFAs and to capture economies of scale, companies have sought to increase the size of their concessions (to as much as 600,000 hectares) by acquiring more than one UFA and associating with or acquiring other companies, includ-

Table 5.2. Timber concessions by area, firm, and nationality of main shareholders

Group name	Companies	Nationality of main shareholders	Area under concession (ha)	Share of allocated areas (%)
Thanry/Vicwood	CIBC, CFC, SEBC, SAB, SEBC, J.Prenant, Kieffer	China/Cameroon	663,288	12
FIP-CAM	FIP-CAM	Italy	539,016	10
Rougier	Lorema, SFID, SOCIB, Cambois	France	474,164	8
SEFAC	Filière bois, SEFAC, SEBAC	Italy	411,872	7
Khoury	EFMK, SABM SN COCAM, RC Coron	Lebanon/Cameroon	338,092	6
Alpi	ALPICAM, STBK GRUMCAM	Italy	305,000	5
Pasquet	Pallisco, Assene Nkou Sodetrancam	France/Cameroon	301,387	5
Wijma	COFA/WIJMA	Netherlands-Cameroon	242,021	4
SIM	SIM, SCTBC, SFDB, INC Sarl	Italy	210,470	4
Patrice-Bois	COFA, SFF, Patrice Bois	Italy/France/Cameroon	206,866	4
Others			2,608,353	36
Total			5,721,583	

Source: GFW 2006.

Figure 5.8. Timber exports by firm (roundwood and roundwood equivalent), 2003

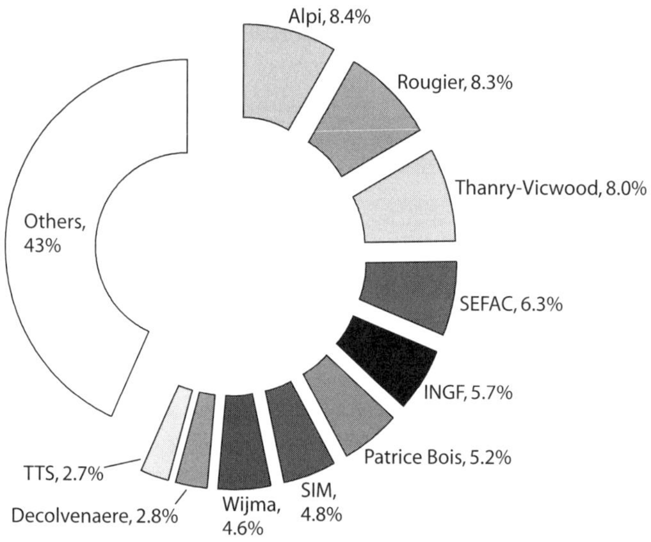

Alpi, 8.4%

Rougier, 8.3%

Thanry-Vicwood, 8.0%

Others, 43%

SEFAC, 6.3%

INGF, 5.7%

Patrice Bois, 5.2%

TTS, 2.7%

Decolvenaere, 2.8%

Wijma, 4.6%

SIM, 4.8%

Source: Authors, based on GFW Atlas, 2004.

ing those with processing facilities (table 5.2). In the early 1990s, most of the companies mentioned in table 5.2 were independent; by 2004, about 65 percent of the concession area was controlled by fewer than 10 groups. The trend toward concentration is also significant for the export industry, given that the seven largest industrial groups accounted for 44 percent of wood exports in 2003 (figure 5.8).

The increase in area controlled by industrial groups did not undermine competition or favor oligopolistic practices. In fact, consolidation was accompanied by diversification in the origin of capital, company ownership, products, and markets. The French and Lebanese presence shrank as new investors arrived. Vicwood, a firm based in Hong Kong, China, acquired the Cameroonian subsidiary of the French group Thanry, giving China a significant share of Cameroon's forest concession area. Malaysian groups subcontracted with domestic concession holders following the devaluation but then, reluctant to participate in the competitive process for acquiring harvesting rights, left the scene. The presence of Italian companies increased in Cameroon because of the log export ban. A typical case is PLACAM, a northern Italian company that specializes in ply veneer of ayous and invested directly in Cameroon in response to the anticipated ban.

Three large local companies (Ingeniérie Forestière, STBK, and SCTB) have emerged and invested both in new concessions and processing equipment, while two domestic companies (SEFICAM and SETRABOCAM) have invested mostly in wood processing without possessing forest harvesting rights. Some Cameroonian firms have acquired concessions competitively and created new processing plants, although most continue to operate through partnerships with larger, often foreign, investors.

The diversification of capital, the arrival of new investors, changes in concession ownership, new acquisitions of concessions, the demise of some of the old timber barons, and the affirmation of national companies are all generally positive developments. This diversified forest industry seems less focused on maintaining old privileges and more open to change. Concentration—a typical response in a sector facing more costly access to resources in the midst of more demanding social, environmental, and legal obligations—has created large companies capable of setting new social, fiscal, and environmental standards for the sector, which smaller companies are now starting to follow. Concentration has not prevented new, smaller companies from emerging in the formal and informal sectors.

The resulting, more diversified structure has strengthened the forest sector. Vertically integrated enterprises provide a sound industrial basis and steady fiscal revenues. Smaller and informal enterprises supplying local markets have contributed to vibrant job creation in rural areas. The smaller enterprises have developed mostly outside the law and must be integrated into the formal sector, yet their presence is extremely important to foster balanced development in the industry. Favoring the bigger formal sector against the smaller informal sector, or going to the other extreme, would be bad policy.

Changes in the processing industry.   The 1994 Forest Law—which required 70 percent of wood to be processed domestically—and the 1999 partial ban on log exports caused industry to expand processing capacity. That capacity rose from 1.2 million cubic meters in 1994 to 2.3 million in 2004, peaking at about 2.6 million cubic meters in 2000/01 (CIRAD 2000; Fochivé 2005).[16] Logs, which represented 57 percent of Cameroon's timber exports by volume in 1998/99, accounted for about 5 percent in 2004 (box 5.4). Sawn timber rose to 85 percent of exported timber volume in the same period. More processed products, such as veneer, plywood, and by-products, increased less rapidly and currently account for about 10 percent of exports.

New processing capacity was installed in response to the law rather than in an effort to improve technical or economic efficiency. France's Agricultural Research Centre for International Development (CIRAD)

Box 5.4

What happened to Cameroon's log exports following the ban?

To develop local processing capacity and encourage the marketing of lesser-known tree species, Cameroon made it illegal to exports logs of the main harvested woods in 1999. Exports of ayous (the most harvested type of wood), azobé, and lesser-known kinds were still allowed, but a surtax was charged and quotas were sometimes imposed. The exception for ayous (and to a lesser extent azobé) was designed to soften the transition for the private sector as well as for the Treasury, given that fewer exported logs meant less tax revenue.

After an initial sharp drop, the volume of log exports more than doubled between 2005 and 2006 following changes in the relative prices of logs and sawn wood (logs were in high demand in the thriving Asian market). In addition, some formerly little-used woods such as okan are exported in larger quantities without any restriction other than export taxes. Similar trends are seen in other Congo Basin countries. Cameroon is likely to discourage any return to an export structure dominated by log exports (either through higher taxes or quotas), given the implications for reducing employment in the wood-processing industry.

(2006) found that investments were modest from 2000 to 2006, because less financially sound companies were unable to invest in new technology as new and more demanding regulations came into force. This trend has accentuated the divide between companies that are backed by international groups and those that are not.

A 2004 inventory listed 44 wood-processing units with a total capacity of 2.3 million cubic meters (Fochivé 2005) (appendix 8). Although capacity in 2004 approached capacity in 1998, dramatic changes had occurred in the intervening period: Seventeen sawmills, with a processing capacity of 400,000 cubic meters, were closed owing to severe log shortages, rising production costs, technical inadequacy, or administrative sanctions. Fifteen new processing units, including five sliced-veneer and veneer/plywood units, were set up, which represented a broad technical upgrade and expansion of processing capacity. By 2005, another eight mills had closed because of industry restructuring. Aggregate processing capacity for Cameroon's forest industry is now estimated at 1.9 million cubic meters.

Because excess processing capacity often contributes to illegal logging in tropical forest countries, it is vital to monitor.[17] The most recent data (Fochivé 2005) found installed capacity to be 300,000 cubic meters in excess of estimated sustainable harvests of major timber species.[18] Although surplus capacity appears to offer incentives for a processor to seek illegal sources of timber, it is important to recognize that processors do not necessarily choose to operate at 100 percent capacity all of the time. Cameroon still exports 150,000–200,000 cubic meters of logs annually. A company may decide to operate mills at or under full capacity depending on international prices and its marketing strategy for processed wood versus logs. The government still issues rights to ventes de coupe (about 60 are harvested every year), whose production complements that of UFAs. Additionally, some potential may exist for timber-deficit companies to acquire timber from companies whose supply exceeds their processing capacity. A comparison of each company's raw material requirements and supplies of timber from UFAs under its control indicates that 35 Cameroon-based companies lack sufficient raw material from their UFA concessions and must procure it from external sources, and 34 have a surplus of raw material.[19] Among the industrial groups, 2 had potential surpluses and 17 had potential deficits. Adjustments are thus still under way in UFA control, company ownership, installed capacity, and timber sourcing. Local trade between companies may become more common.

Employment in the industry.   Despite a reduction in the volume of timber harvested between 1998 and 2004, the number of employees in the industry increased (figure 5.9). Though relatively modest, this increase conceals more significant changes:

- Employment in logging has declined by 20 percent, and employment in processing has risen by 25 percent.
- Processing units have relocated from rural to urban areas to attract a workforce that is more qualified to run more sophisticated processing units. At the same time, smaller and informal enterprises supplying local markets have fostered employment in rural areas (see the next section).
- The nature of the workforce is changing. Companies that formerly hired only mechanical engineers and technicians to manage logging operations have started to hire professional foresters to develop and implement management plans.

Figure 5.9. Employment in logging and processing, 1998 and 2004

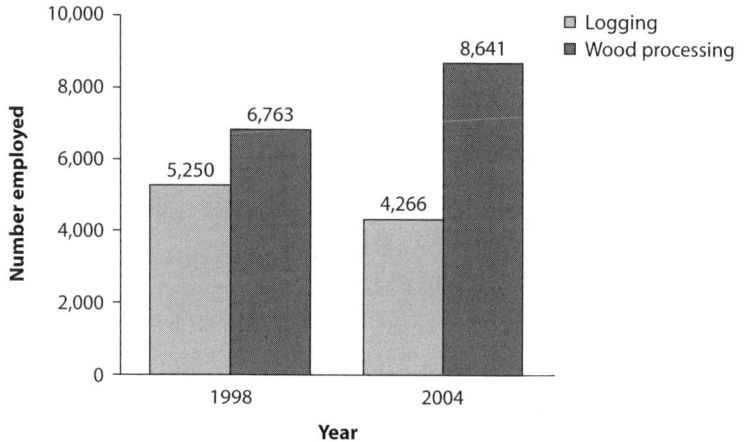

Source: Carret, Giraud, and Lazarus 1999; Fochivé 2005.

Trends in artisanal forestry.   As mentioned, artisanal forestry started to expand after the 1994 devaluation, when industry abandoned domestic markets, and grew further after 2002, when greater numbers of community forests were developed. In the absence of policy and legislation regulating small-scale artisanal harvesting and processing, these activities are mostly informal, not mandated to abide by specific forest rules, and often associated with illegal logging. The simultaneous rise of artisanal logging and community forests was unfortunate because the community forestry banner was often used as a cover for illegal logging.

In artisanal logging, timber is generally cut in the forest using mobile saws. Based on an estimate of mobile saws sold in Cameroon, it would appear that 4,800 artisans engage in timber production. Figures provided by Cameroon's customs service in 2004 indicated that 18 artisanal companies were registered as exporters for 138,098 cubic meters (one-fifth) of sawn wood exported in that year. In other words, artisanal logging was supplying not only domestic but international markets and competing unfairly with formal exporters, which had to pay taxes and submit to other industry regulations.[20] The Ministry of Finance decided to levy stiff fines to discourage exports of artisanal timber,[21] but the measure was not well enforced, caused trouble for all exporters, and had modest impact on limiting illegal timber exports.

Effects of Fiscal Reform

The increase in revenues under the new forest taxation system was discussed in chapter 3; this analysis examines the effects of fiscal reform on

Table 5.3. Taxes on major unprocessed and processed wood products

Taxation	1992		1998		2004	
	Ayous	Sapelli	Ayous	Sapelli	Ayous	Sapelli
On log (m³)	5,094	11,344	20,650	28,725	33,047	41,536
On sawn wood (m³)	—	4,073	29,848	37,953	40,242	25,733
On veneer (m³)	—	2,444	21,946	28,828	24,145	18,349
As percent FOB value for logs	10.7	15.1	25.6	23.7	36.7	32.4
As percent FOB value for sawn wood	—	2.9	14.7	13.0	19.2	15.5
As percent FOB value for veneer	—	2.2	8.1	10.7	8.9	6.8

Source: CIRAD 2006.

Note: Calculations based on ayous and sapelli, the two major woods that Cameroon harvests and exports.

— = not available.

industry, for which the tax burden increased, with the implications for investment and competition mentioned earlier. To gain some idea of the effects of fiscal reform over time, forest taxation was compared at three significant times: in 1992, prior to the 1994 Forest Law; in 1998, before the auction system was implemented and higher area taxes were levied; and after 2004, when reforms had been more fully implemented. In determining the impact of fiscal reforms, parameters other than taxation were assumed to be constant over time: average harvest intensity (10 cubic meters per hectare per year); duration of harvesting cycle (30 years); and processing yield (33 percent for lumber; 55 percent for veneers). For additional details on the methodology and results, see appendix 9.

The tax burden has increased significantly for all types of wood products since 1992 (table 5.3). This trend has been particularly marked among processed products, which were virtually untaxed in 1992. The tax on logs remains the most important tax per cubic meter of product, reflecting the national priority assigned to domestic timber processing.

A subregional comparison. Timber companies in Cameroon sustain a higher tax burden than their counterparts in other Congo Basin countries (table 5.4). Even though this statement is factually correct (except for logs and veneer from Equatorial Guinea), the variation is not wide. Aside from the tax burden, a number of considerations feature in a firm's calculus about where to operate, including transport costs and the "cost of doing business"[22] (higher in other countries in the region), the size of

Table 5.4. Fiscal burden for exported wood products in Cameroon and neighboring countries, 2005
Euros per cubic meter

	Cameroon	Congo, Rep. of (North)	Gabon With MP	Gabon Without MP	Central African Republic	Congo, Dem. Rep. of	Equatorial Guinea
Estimated fiscal burden for exported products (€/m³)							
Logs	40	30–52 (14–24)	39 (26)	41 (27)	40 (19)	28 (14)	62 (41)
Sawn wood	49	42 (8)	30 (10)	35 (12)	53 (8)	35 (7)	
Veneer	29		19 (5)	22 (6)			47 (12)
Average transportation cost (m³ log to port)	40–70 (remote Eastern regions)	105–110	10–45		105–110	20–130 (main areas 50 to 80)	10–30

Source: CIRAD 2006.

Note: The numbers in parentheses show the tax as a share of the free-on-board price.

concession, and value of its tree species (also higher in other countries in the region).

Impacts on Poverty and Livelihoods

Forest reforms had a number of effects related to poverty and livelihoods. Experience with the evolution of indigenous peoples' rights, community forestry, and local timber markets under the reforms provides substantial guidance to develop and test alternatives that may offer better means of sustaining forest communities and the resources on which they depend.

Improving Indigenous Peoples' Access to Forest Land and Products

About 30,000 individuals—Aka, Baka, and Kola peoples gathered in or around 300 settlements in 33 councils—constitute Cameroon's indigenous peoples. Legally, they are citizens with rights equal to those of all other citizens. Yet they lack the de facto political influence; legal status; and organizational, technical, and economic capacity enjoyed by other groups. Fundamentally, their settlements are not recognized "communities," and therefore they cannot interact legally with government services. Although they are the most forest-dependent people in Cameroon, they have no legally sanctioned access to the forests where hunting and gathering generate more than 65 percent of their livelihood, on average.[23]

The 1994 Forest Law began restoring some of the rights that colonial and postcolonial rule had taken from forest people. Although much remains to be done, the reforms provided a much-needed entry point for improving the livelihoods of indigenous people (and indeed of all people who rely on customary access to forests) (box 5.5).

The experience with Cameroon's forest reforms revealed some limitations in helping indigenous people secure rights to their forest resources. First, forest zoning allocated a significant area of indigenous community land to the permanent forest domain, where community forests cannot be established. The current regulatory framework for these areas does not sufficiently specify local inhabitants' rights to hunt, gather, or fish. Second, in the nonpermanent estate, where community forests *can* be established, the dominant Bantu communities have already claimed the land, perpetuating a history of marginalization of Pygmy people.[24] Finally, aspects of community forests are inconsistent with the livelihoods and resources of indigenous people, such as the emphasis on small-scale timber production or administrative structures that conflict with traditional forms of land use.

Box 5.5

Rights and achievements of local populations following forest reform

Cameroon's 1994 Forest Law recognizes the existence and use of traditional forest rights, which are the rights that people traditionally living near or within forest areas may exercise with a view to satisfying their need for forest products. These rights are free of charge and freely accessible, as long as the beneficiaries maintain geographic proximity to the forest, harm no protected species, and remove forest products only to meet their personal or collective and strictly noncommercial needs. The Forest Law stipulates, however, that these rights may be restricted or even entirely revoked if they become incompatible with sustainable management of forest resources.

Recognition of these rights is one of the most visible and tangible changes brought about by the decentralization of forest management. Local populations—including indigenous people—may use forests even in protected areas such as Lobéké in southeastern Cameroon and Campo Ma'an in the south. They participate in forest zoning and have set limits on areas allocated to commercial timber concessions (for example, UFA 004 allocated to TRC in the Ebo'o forest). They have established community forests and community hunting zones. A significant share of the revenue from forests and wildlife is dedicated to the development of the surrounding communities, which increasingly participate in decisions related to forest management.

Source: Adapted from Bigombe Logo 2008.

In the absence of appropriate measures to protect their rights, forest interventions—running the gamut from illegal logging to biodiversity conservation—threaten to increase the marginalization and impoverishment of Cameroon's indigenous groups by depriving them of traditional forest use rights with little or no compensation. Potential ways of overcoming these limitations are discussed in the next chapter.

Do Forest Revenues Reach Communities?
Because communities do not enjoy legal status and cannot receive public money directly, the mayors (locally elected officials) who head the local councils are the primary recipients of the shares of area tax corresponding to councils and communities. Indigenous groups, because they not

only lack legal status but are marginalized in many respects, are generally excluded.

Since 2000, about $12 million per year has been transferred to a population of about 3.2 million people represented by 56 local councils. The 2004 audit of $53 million in tax revenues transferred to the local level over five years showed that the lion's share of these funds (exceeding 65 percent of both investments and operating costs) was used to improve the infrastructure and operation of local councils themselves (table 5.5).

How well do the mayors' priorities for using forest revenues reflect community priorities? Villagers surveyed about their priorities emphasized the isolation of their villages, the difficulty of obtaining credit, the volatile prices of local products, the lack of social infrastructure (such as medical centers), and the lack of housing. In fact, investments in the development and operation of community services (roads, education, health, and sanitation) were made only when forest revenues were substantial and exceeded the immediate needs of local councils. Even though the audit revealed that resources allocated to community services increased between 2000 and 2004, the modest level of these investments (see table 5.5) apparently fueled the sentiment that mayors and forest management committees were appropriating most of the revenue. Given that it is virtually impossible to visibly link the removal of trees with the eventual arrival of funds and that little trust exists between the governed and their governing bodies, local perceptions of the misuse of funds are hardly surprising.

There is some potential to address the shortcomings of these financial transfers, although it is virtually impossible to keep the community forest revenue stream completely separate from other revenue flowing to the mayors. The economic and financial audit of the forest sector conducted at the government's request[25] in 2006 suggested that forest revenues could be redistributed more efficiently to alleviate rural poverty by capping the yearly revenues that any local council can receive at $160,000 or 200,000 (CFAF 80 million or 100 million); using the remaining resources to increase revenues for local councils with modest and declining forest revenue; and enabling local councils near protected areas (and not only those near concessions) to receive forest revenues. It has also been suggested that insitutional reform could foster the development of truly representative local institutions to receive the funds, versus sending them to institutions somewhat removed from the communities themselves.[26] Such reforms could also include formal and systematic representation of communities and indigenous people on committees that manage these

Table 5.5. Use of forest revenues at the local level, 2000–04 (overlapping council and community priorities are shaded)

Area tax redistribution and use	Local councils	Communities	Total
Inflows (redistribution amounts, $000)			
East Province	22,662	5,671	28,333
South Province	7,212	1,806	9,018
Center Province	6,154	1,542	7,696
Southeast Province	1,196	299	1,494
Coast Province	630	158	788
West Province	16	4	20
Total	37,870	9,480	47,349
Outflows (expenditures, in %)			
Investment in:	55% of redistributed revenues	72% of redistributed revenues	
Roads	13	7	Roads
Health, education, market infrastructure	5	20	Health
Local council investments (building, housing of representatives, transport, operations, and management)	66	17	Culture and sports
Other	19	14	Water distribution and sanitation
Total	100	13	Housing
		9	Social
		7	Creating community forests
		13	Other
		100	Total

revenues and a national directive mandating that all populations, including indigenous populations, benefit from these transfers (Bigombo Logo 2008). A new mechanism is being developed to fully integrate decentralization, taxation, and local revenue allocation.

Such measures would enable forest revenue to contribute more fully and appropriately to decentralization and poverty alleviation, and they would also help communities move beyond acrimonious debates over equity, transparency, and profiteering. At this initial stage, elected officials

Table 5.5 continued

Area tax redistribution and use	Local councils	Communities	Total
Recurrent	40% of redistributed revenues	20% of redistributed revenues	
Education	14	36	Education (teachers' salaries)
Health	3	2	Health (salaries)
Local council operational costs (meetings, allowances, transport, operations, and mangement)	42	42	Community financial audit
Local council salaries	25	20	Services for creation of community forests
Culture and sport	4	100	Total
Other	12		
Total	100		

Source: Institutions et Développement, November 2004 update of audit on the redistribution of forest taxation revenues.

are using forest resources to finance their offices and programs. Community interests come second. Can this divergence of interests be resolved? It is expected that local governing bodies will come to express the will of citizens over time as democratic structures consolidate. The question that remains is whether potential conflicts between local, traditional forest resource authorities (elders, lineage leaders, village chiefs), and centrally imposed institutions can be resolved in ways that support true community action and representation with respect to forest management and revenue.

Community Forests and Local Livelihoods

Observers concur that community forestry in Cameroon is several years ahead of similar efforts in other parts of Africa's humid tropics and that lessons from Cameroon will be useful to other countries (Julve and others 2007). Communities can now exert their rights over forests by establishing community forests, by preempting forest areas from being tendered for ventes de coupe, and by converting forest land to other uses.

These rights are still unevenly and imperfectly enjoyed, however. Several approaches have been attempted, and the number of community forests has grown since 2000, but no single approach to community forestry has yet delivered the entire set of anticipated benefits. A 2006 review of community forests led by national and international NGOs highlighted the wide disparity between outcomes envisioned in the 1994 Forest Law and actual outcomes in the field, where neither poverty reduction nor environmentally sound management goals had been met. The bottom line is that, to date, Cameroon's few successful community forests have been heavily supported and monitored by donors, and their sustainability remains uncertain. Community forest revenues are often modest and do not always reach the communities themselves. A cumbersome regulatory framework and a welter of social, economic, technical, and other constraints at the local level make community forestry almost unworkable.

Ultimately, community forestry has a lot in common with other community initiatives. Some regard it as naïve to expect that it can thrive in a country where participation is only starting, administrators use public office for personal gain, vested interests prevail, local governing bodies are dysfunctional and starved for funds, and systems of traditional rights and obligations are on the wane. Others believe that community forestry would work if reforms had followed a better sequence (box 5.6). Whichever stance is taken, the community forestry experience offers several considerations for devising a more workable approach.

"Communities" in the rainforest zone may not fit outsiders' assumptions about communities and how they function.   Cameroon's "communities" are not necessarily stable, homogeneous social entities (Burnham 2000; Malleson 2001; Biesbrouck 2002; Geschiere 2004). Contrary to common assumptions, the land surrounding villages is traditionally claimed by families, not by communities (Geschiere 2004), many of which coalesced when colonial authorities brought people to live alongside roads to provide hard labor for road maintenance (Takforyan 2001). These composite communities may not be the kind of resilient social institution that is so often visualized by outsiders as the backbone of community forestry. One of the more intractable problems to be resolved, highlighted by many long-term observers, is the poor governance that arises in communities because of conflicts between community members and elites of different ethnic origins and allegiance.

Tenurial issues add complexity.   Tenurial arrangements that apply to trees can differ depending on where the trees are located (Vermeulen and Carrière 2001). Boundaries of "community" land are rarely defined

Box 5.6

Community forest reforms: Too little, too late?

Community forestry is one of the key areas of Cameroon's forest sector reforms. Community forestry schemes should enable communities to participate in decisions and draw more economic benefits from the forest. Yet community forestry initially received less attention from the international community and government of Cameroon than efforts to regulate the formal forest sector and restrain corrupt and illegal practices.

With time, a stronger national and international constituency has emerged to advocate for community forestry, especially to support social welfare goals, forest management that reduces environmental impacts, and environmental services. Currently community forests are near the top of the forest agenda and receive considerable attention from donor agencies.

Has this attention come too late? Could stronger, earlier efforts to establish community forestry have stemmed the tide of illegal logging by paying more attention to legalizing communities' forest rights, establishing more benign partnerships with industry than those that have emerged, and ultimately benefiting communities more than community elites?

It is not clear that community forestry could have developed and been adopted more widely without bringing the expanding weight of industrial forestry and political patronage under control. Throughout the forest-rich countries of Central Africa in the 1990s, forest laws, tenurial regimes for land and trees, and Forest Administration mandates gave industry a highly privileged place in relation to local populations and the environment. Under these circumstances, regulating industrial forestry and creating instruments that would strengthen people's claim on forests appeared to be a prerequisite for promoting community forestry in Cameroon. Even in hindsight it remains doubtful that community forestry strategies would have gained ground otherwise.

with precision unless natural boundaries such as rivers are present, and conflicts are frequent when boundaries are set (Takforyan 2001; Karsenty, Mendouga, and Pénelon 1997). Most community forests are created in areas that lie between two villages and are claimed by both—perhaps indicating that the desire to establish a community forest is based largely on the desire to protect land from the incursions of others.

It is increasingly common for categories of forest tenure to overlap. For example, a timber company can have rights to harvest a production forest where rural communities have traditionally gathered forest products. Depending on how these rights are handled among the various stakeholders, the result can be conflict (over hunting or for gathering of some nontimber products associated with commercial trees) or peaceful coexistence and mutual benefit.

Bureaucratic and regulatory barriers remain.   The path to creating a community forest remains challenging, slow, and costly to travel. The Forest Administration intrudes excessively in local decision making and management. Requests to create a community forest can take from 1.5 to 5 years to study and approve, and the subsequent cost of developing and gaining approval of a management plan is expensive, ranging from $2,800 to $32,000 (CFAF 1.4 million to 16 million) (Lescuyer, Ngoumou Mbarga, and Bigombe Logo 2008). The inventory of forest resources requires a high level of technical skill, and the results are rarely used in developing the simple management plan. Thus, no community can establish its rights to a forest without technical and financial assistance from an NGO or some other external entity, raising the risk that these external actors will take control of the process away from communities. Far from encouraging sustainable forest management, these deficiencies have encouraged the spread of illegal activity in community forests.

In some cases, the challenge of finding 5,000 hectares of contiguous forest land has led neighboring villages to present joint applications. This approach facilitates the creation of community forests, but disputes between villages are common and the risk of breaking agreements high.

Generating significant income from community forests has proven difficult.   Whether communities decide to focus on timber or on other forest products, it is proving difficult for them to draw a significant livelihood (legally) from community forestry.

Communities are free to manage their forests for a wide variety of purposes. Communities may decide not to exploit timber and focus on nontimber forest products, but they rarely do so. Outside the narrow boundaries of the community forest (in rural areas and forest concessions), communities can gather large quantities of products other than timber, either freely or with easily obtained permission, when this activity is compatible with using the forest to harvest wood. In other words, most nontimber forest products do not need community forests to grow or to be harvested legally, and few communities see the need to clear numerous bureaucratic hurdles merely to formalize activities they have pursued for countless generations. Community forestry based on nontimber resources has been successful elsewhere (Brazil nuts, harvested by forest communities in South America, are one example). Similar options may emerge in Cameroon, but until then, a focus on timber appears more likely.

Communities opt for logging because timber appears to generate revenues higher than those of other forest products—a situation that encourages a dependence on industry that will not be viable over the long term. Lacking a market culture, communities often enter into poorly developed contracts or operate without a contract, which can be a source of conflict with loggers and buyers. Yet as long as timber remains their main economic focus, community forests cannot offer a realistic alternative to industrial concessions. Forest operations require expensive equipment to cut and transport timber and skills to run companies and access remunerative markets. There seems to be potential for synergy in operating industrial and community forests and generating steadier and higher incomes from community forests. Even this option cannot work without substantial regulation, oversight, and equitable agreements between industry and local communities.

Since 2002, timber operations in community forests can be done only with simple implements and chain saws. In the absence of outside help, the production is of limited value and the potential for trade limited. Given the capital and logistical constraints, it is entirely possible that community forests relying on small, artisanal timber enterprises are simply not viable in entire areas of the country, especially the more remote and sparsely populated ones.

Collusion and corruption may prove more enduring than any other constraint to community forestry. Cuny and others (2004) have concluded that the notion of establishing a community forest "rarely comes from the communities, but rather from all the other actors who have an advantage in the process." Community forests often originate through the self-interest of prominent village-originated individuals with the social capital to gain authorization for a community forest; sometimes these "elites" are urban civil servants who visit the village only rarely (Oyono 2005; Bigombe Logo 2008). In any case, they do not represent the full spectrum of village interests. Rather they are generally well positioned to act as middlemen between the village and industrial interests, which may use the community forest to avoid the regulations and especially the taxes imposed on UFAs (for example, timber harvested in community forests is not subject to area taxes or felling taxes).

The self-interest of elites is symptomatic of wider problems. Cuny and others (2004) have observed that collusion is especially detrimental to easing poverty, because it excludes communities from the formulation of management plans and benefit sharing. Government reports confirm that "in most community forests . . . operations [including] technical exploitation and marketing of wood are monopolized by economic operators who have signed partnership agreements with communities [and] members of the community are being left out of the process. Also, all operations to be carried out in the forest become commercialized."[27]

Local Council Forests

The gazetting of local council forests, like the gazetting of community forests, remains a costly and time-consuming process. It must be undertaken by the local council itself. It appears that many mayors simply prefer to obtain the percentage of the area tax corresponding to the portion of the UFA that overlaps local council boundaries.

Local council forests have been criticized severely for the misappropriation of revenue by mayors and their clans. Villages under the jurisdiction of a local council may not be treated equally with respect to investments and sharing of tax revenue. These dysfunctional behaviors are not unexpected. The only alternative is for citizens to confront these accountability issues with all of the legal means they can muster (with the help of foreign partners and allies). This alternative may function somewhat better in an increasingly democratizing context, in which an independent press, the Internet, the judiciary system, and newly created institutions such as the Court des Comptes provide an opening for greater public discourse and accountability.

There are emerging signs that the use of forest income may be improving in some communities. In a very small sample of villages in southern Cameroon, Lescuyer, Ngoumou Mbarga, and Bigombe Logo (2008) found that "the previously held negative assessment" of misuse of forest income "is not universal" and that "increasing numbers of villages . . . are making good use of this income." They recommend that further work be done to test these findings, and suggest that the redistribution of forest revenues will improve if communities can receive funds directly and become full partners in managing them, and if the entire spectrum of village interests (and not, for example, those of a single clan) is represented.

Impacts on Rural Markets and Small Enterprises

Local markets and small-scale enterprises are the source of numerous jobs for the poor—precarious but essential—yet little effort was made to assess the potential impacts of reforms on local employment, supply chains for the local timber market, and revenues. For example, Cameroon suspended small titles, which were notoriously difficult to monitor, in 1999. This move was supported widely by the international community to curb illegal logging by industry, and industry generally complied. This action, while justified, caused a shortage of building material in some rural markets and reduced employment. As noted, small titles are being issued once again, following new directives in 2007 that made them somewhat more amenable to monitoring.

Notes

1. Details on the methodology and results are available in the published paper. It appears that the 50 percent figure does not derive from any scientific analysis but rather from an informal assessment, given that no method or source was cited for obtaining this result. The figure was then widely—and improperly—disseminated by many organizations (Cerutti and Tacconi 2006).
2. In one well-known instance, about 50,000 cubic meters were estimated to have been harvested illegally from UFAs 10-030 and 10-029. Though the violations were well documented by the independent observer, the court case is still pending.
3. *"Au centre des problèmes de gouvernance au sein du MINFOF."*
4. It should be noted, however, that the overall decline in illegal timber exports from managed concessions has had a large impact on the overall reduction in illegal logging throughout Cameroon.
5. Many taxes (such as mill-gate taxes) are probably not paid, but this does not mean that the logs come from sources not covered by any permit.
6. Since 2006, wildlife infractions (poaching) are also part of the press release on illegal activities.
7. A major case of illegal logging, awaiting resolution in court, accounts for fines of CFAF 15 billion.
8. See REM (2007). Most fines decided in this way in 2005–06 were reduced by 70–98 percent from the initial amount.
9. 0.28 percent and 0.14 percent are gross deforestation rates; these rates are not affected by reforestation of deforested areas.

10. WWF used a metric known as "protected area management effectiveness" (PAME) and scored each area using the PAME Tracking Tool, which is based on 30 indicators that provide a comprehensive assessment of management effectiveness. The tool, developed jointly by WWF and the World Bank, is consistent with management effectiveness recommendations of the World Commission on Protected Areas and with the Global Environment Facility's monitoring and evaluation policies.

11. The 14 other UFAs were allocated in 2005 and 2006 and are still under the provisional management convention.

12. This figure is consistent with the 6 million hectares of production forest where harvesting rights have been granted, an average harvest volume of 10 cubic meters per hectare, and a 30-year felling cycle (which would yield 2 million cubic meters). The 500,000 cubic meter difference supposedly comes from ventes de coupe, community forests, and small titles.

13. Small titles were suspended by a Ministerial Decision from 1999 to 2006.

14. Studies of harvesting intensity have also indicated that it is very location specific, depending on such factors as concession holders' habits and markets or the distance to ports and local markets. The situation of a rich, intensively managed UFA with harvests of 20 cubic meters per hectare (including 55 percent sapelli and 29 percent tali) differs very significantly from a community forest providing 30 cubic meters per hectare of annual coupe to a local urban market.

15. With the exception of a handful of experienced wood processors, such as Alpicam and ECAM-Placages.

16. This figure does not include the additional capacity of mobile (Lucas Mill) sawmills, estimated at about 300,000 cubic meters (Fochivé 2005).

17. The CERNA-ONFI-ERE development study (Abt and others 2002) estimated a supply gap of 1.2 million cubic meters.

18. The estimated processing capacity is 2.3 million cubic meters. Based on 200,000 hectares annually available for selective harvesting within 96 UFAs occupying 6 million hectares (9 UFAs being conservation concessions) and a harvest intensity of 10 cubic meters per hectare per year, the annual production of UFAs is estimated to be 2 million cubic meters, and surplus processing capacity is estimated at 300,000 cubic meters.

19. Updating the previous exercise by Abt and others (2002).

20. Artisanal timber is exempt from most taxes, such as the area tax and sawmill entry tax.

21. To this end, the 2005 Finance Law (Article 247 bis) mandated that all sawn wood arriving at the port with insufficient proof of legal origin be submitted to a penalty of 400 percent of the equivalent of all taxes, including area fees, felling tax, and sawmill entry tax.

22. The "cost of doing business" is not necessarily the burden imposed by corruption but also refers to more or less legal levies on economic activity, such as costly *cahiers des charges*, social responsibility contracts obligating companies to in-kind and cash transfers to local communities, and (at times) transport, per diem, or other expenses for local civil servants.

23. "People living within or near Cameroon's forests have traditionally used the forest for hunting, fishing, gathering, agriculture, and even for raising small livestock—activities that are central to food security. The forest is also a spiritual resource, because it serves as the link between the living and the dead and is the place where rituals and traditions are enacted. These two dimensions—the material and spiritual—transform the forest into a space that is far more than a storehouse for timber" (Bigombe Logo 2008).

24. With the notable exception of the Moangue–Le Bosquet community forest, located near Lomié and run by sedentarized Baka Pygmies.

25. The government showed little interest in the results of the audit; an official version of the document has never been published.

26. Arrêté 122 would have to be revised to permit such institutions to be created, a suggestion that has been received favorably by the Ministry of Finance and the Ministry of Territorial Administration.

27. *"Appropriation du processus : (...) dans la plupart des FC (...) le promoteur s'en arroge la paternité ou (...) les opérations techniques d'exploitation et de commercialisation du bois sont monopolisées par les opérateurs économiques ayant signé des conventions de partenariat avec les communautés; la conséquence en est que les membres de la communauté concernée sont mis à l'écart du processus. Aussi, toutes les opérations à effectuer dans la forêt deviennent payantes."*

PART III

Conclusion

CHAPTER 6

Ten Years of Forest Sector Reform

Although the lessons of the reforms will continue to unfold, 10 years is an appropriate juncture to review what has been learned and identify gaps in knowledge and action. This chapter discusses the lessons learned in developing and implementing reforms and identifies elements of the reforms that require continuing attention.

Lessons for Developing and Implementing Reforms

The findings and lessons described below can be divided into two categories: those that may be relevant to comparable efforts in neighboring countries of the Congo Basin and beyond; and others offered as additional considerations to the government of Cameroon and civil society as they develop their vision for managing and using forests to improve and sustain local livelihoods.

There Is No Reform without Reformers—At All Levels

In a country where political interests so often take precedence over the public interest, the presence of people who are willing and able to change the status quo must never be taken for granted. Reforms would certainly not have moved forward without support from a group of reform-minded officials within Cameroon's Technical Oversight Committee for Economic Reforms (Comité Technique de Suivi des Réformes Economiques).[1] As support increased over time, the reformers moved beyond the initial measures—for example, by securing the presence of independent observers at the auctions and in pursuing harvesting infractions.

A parallel lesson is that the support of reformers and champions does not last forever. It can weaken or vanish as political, personal, and economic priorities change. Support from the Technical Oversight Committee and other prominent reformers has faltered in recent years for a number of reasons, probably ranging from the personal to the political and economic.

At the same time, reliance on influential reformers can catalyze reform but not complete it. It is essential to foster awareness and support for reforms throughout civil society, industry, and all branches and levels of government. After 2000, reformers made an active effort to engage other actors, including members of parliament, national NGOs, and private organizations. A coordinated approach to forest issues was adopted by the World Bank and other donor agencies. By 2004, members of parliament who were especially knowledgeable about forest issues formed the core of a subregional organization, the Central African Network of Parliamentarians for Forest Ecosystem Management. These members prompted parliamentary debates and engaged the ministers of finance and forests on subjects such as corruption, law enforcement, and national budgetary allocations for forest sector development.

Reform Takes Time and Pragmatism

It was important to be realistic about the time needed to build ownership for forest reforms in Cameroon and to adopt a pragmatic approach that balanced consistency with flexibility. Time was needed for government, public opinion, and the private sector to learn about, assimilate, and ultimately validate reform. The reform agenda (the matrix described in chapter 2) was publicized and put at the center of discussions with government, donor partners, and NGOs. Even among external observers, it became widely known and easy to follow.

Risk is always involved in introducing reforms. Some of the forest sector reforms had no precedent elsewhere, and they were introduced in a setting widely acknowledged to be challenging. Time was needed to build capacity to learn and to adjust the institutional innovations and regulatory structures to be more consistent with the objectives of reform. In Cameroon, the strong and consistent emphasis on the objective of reforms was balanced by flexibility on how best these could be achieved. When certain SAC III conditionalities proved ill adapted to address the real constraint, pragmatic adjustments were made. For example, less technically comprehensive forest management guidelines were adopted because they were more straightforward to apply under the conditions prevailing in Cameroon. It was understood that policies and regulations were more likely to endure if they were simple and easy to control. The best means of strengthening public sector accountability currently consist of practical measures, such as simplified regulatory frameworks, limited entry points for harassment, and improved public information.

Strengthen Capacity in Forest Institutions

As mentioned in chapter 1, forest institutions are poised to recover from decades of attrition that left them ill prepared to implement new forest policies and regulations. Continued attention to capacity in these institutions at all levels is important. It is essential to ensure that the opportunity provided by high staff turnover at the Ministry of Forests in the next few years is not squandered. New staff must be selected carefully and must have the capacity—in training, credentials, funds, and equipment—to maintain an effective field presence and perform the more difficult regulatory and supervisory functions envisaged by the reform. Poor management and failure to secure adequate salaries and complementary funds will render the next generation of forestry officials susceptible to the corrupt practices of many of their predecessors.

Expand the Community Forestry Framework

The fundamental ingredients for successful community forestry include cohesive communities (which are not always present); a framework that compares well with alternatives (that is, communities achieve benefits that cannot be achieved outside the community forestry framework); a relatively reassuring environment where communities are fairly and democratically run; and public institutions that take all parties' rights and obligations seriously rather than furthering the interests only of administrators and local elites.

The community forestry framework should be expanded beyond the boundaries of conventional community forests to support forest-based, income-generating activities in the broader forest and rural space. Extended areas defined through participatory mapping would create a mosaic of community fields, fallows, and concessions where different private and community uses would coexist based on negotiated agreements.

Detect Problems Early and Support Regulation
with Institutional Innovation

Reforms embody change that occurs in the midst of change. In combination, reforms can have unexpected and undesirable effects. They can erode over time. Their effects can be altered significantly by sometimes subtle changes in the local context. It is essential to pay attention to the signals that problems are developing and deal with them before they become overwhelming. For example:

- Changes in harvesting patterns and intensity can reveal industry's response to new fiscal and regulatory systems. Greater selectivity in the woods harvested can indicate that other factors are preventing industry from supplying a greater diversity of markets despite fiscal and regulatory incentives designed to support diversification.
- Statistics showing that exported volumes of roundwood-equivalent timber exceed total legally authorized volumes, coupled with significant violations of annual authorized harvest volumes and declining domestic timber prices, provide a clear indication of structural overcapacity in the industry.
- When concessions are won by the sole bidder remaining after technical screening, with very low bids, there are probably serious problems with the auction system.
- Other signals allow early detection of problems in supplying local markets, such as increased mobile saw imports reported by the customs service, a shift in sales patterns as reported in forest company annual reports, and the greater size and multiplication of urban timber markets, accompanied by a drop in the price of timber at those markets.

Creating the capacity to detect and act upon such signals is challenging at the beginning of a reform process. It requires determination among reformers and significant external support.

The Cameroon case demonstrates the importance of reinforcing regulatory prescriptions with institutional innovation. Regulatory prescriptions developed under the reforms would have failed had they not been supported by studies and mechanisms that helped identify problems early, especially from the margins rather than the centers of power (for example, by making forest information more publicly and widely available); forums where competing interests could be acknowledged and solutions discussed (such as procedures for communities to claim and adjudicate rights to forest land); and commitment devices (such as the independent observers) to implement or enforce decisions, based on transparent monitoring, feedback, and accountability. Such institutional innovation proved essential for shifting a vicious cycle, in which natural resource depletion and institutional control benefited the few, toward an increasingly virtuous cycle, in which more sustainable management practices and competent institutions enable forest resources to benefit society and the environment.

The Unfinished Agenda

The unfinished agenda is considerable and ever changing. It will need to adapt to new realities as new policy achievements and new policy failures emerge.

Address the Needs of Indigenous Peoples

Special measures are required for indigenous people to participate in and benefit from forest reforms. The government's Indigenous Peoples Development Plan (IPDP) seeks to mitigate risks for indigenous peoples under Cameroon's 2003 Forest and Environment Sector Program, including the loss of control over land that indigenous people traditionally used as a source of livelihoods; loss of the cultural and social identities associated with those lands; increased marginalization; increased dependence on other groups; exclusion from the local, decentralized system of forest administration; reduced assistance from government services; and reduced capacity to defend their legal rights.

Described as the "first major Central African forest policy document specifically addressing indigenous peoples' needs" (Jackson 2004:5), the IPDP envisions a number of corrective actions to give indigenous communities more secure rights to land and resources, especially by giving legal status to indigenous settlements, recognizing their traditional use rights, giving them greater access to markets for forest products, providing new community forests and hunting zones where those rights apply, ensuring greater representation of all marginalized groups within local forest management institutions, and improving indigenous peoples' access to forest revenues for their benefit and that of the land they use (box 6.1) (Jackson 2004; Bigombe Logo 2008).

Give Greater Attention to Impacts on Local Markets, Small Firms, and Employment

International markets may be larger and financially more important than local ones, but continued inattention to local markets has significant social implications and can undermine forest sector performance and governance in many important ways.

As noted, the emergence of Cameroon's informal sector has more to do with the change of relative prices and the progressive impoverishment

Box 6.1

Widening the involvement of Pygmy people in forest management

A promising approach that enables indigenous communities to be legally recognized and manage their affairs in ways that are more in keeping with their own practices and culture is to establish administrative "chieftaincies" (*chefferies*). Such chieftaincies make it possible for Pygmies to gain representation in local councils and parliament. Some Pygmy groups have become involved in forest management by establishing community forests (for example, Moangue Le Bosquet, Nomedjo, and Payo in Lomié in the east and Biboulemam in the south), and they view these forests as a means of securing institutional, political, and social recognition.[a] The secure land rights that accompany the establishment of community forests by Pygmies can support further negotiations over land (facilitated by national and international NGOs) between Pygmies and Bantus under the aegis of the Forest Administration. This approach has been emphasized by the Center for the Environment and Development since 2004, under a project to secure the rights of Bagyéli, Baka, and Bakola people, with technical and financial assistance from the Rainforest Foundation (U.K.), the Forest People Programme, and local nongovernmental organizations. Through such negotiations, Bakola-Bagyéli communities in the districts of Bipindi and Kribi have gained rights to land from their Bantu neighbors, and Bagyéli communities in the south (Akom II District) have created their own chieftaincies. Four communities have already installed community chiefs: Awomo, Mefane, Mingoh, and Nko'omvomba.

As of this writing, 19 Bagyéli communities in Bipindi District have obtained legal recognition of the land that they occupy. Because of variations in local land pressure, these areas range from 0.4 hectares for Log Diga to 1,500 hectares in Bokwi. These results were obtained through negotiations by local government and religious authorities, Bantu and indigenous groups, and NGOs.[b] It is hoped that these achievements, which have been validated by authorities at the very highest level, will endure.

Source: Adapted from Bigombe-Logo 2008.

a. However, relying solely on existing community forest legislation is probably not sufficient, given its current limitations.

b. For a comprehensive analysis of this project's national results, see Handja (2007); Kim (2007); Mefoude 2007; Nelson 2007; and Jackson 2004.

of urban and rural populations than with the regulation of the industrial timber export sector. Given local markets' inability to purchase timber at high prices and their tolerance for modest quality, it is likely that they will continue to be supplied from small, semiformal, and informal companies for some time. Practical measures, including a reduction in the value added tax for local timber sales, should be adopted to ensure that legally sourced timber remains affordable at the local level.

Small-scale artisanal operators, which have emerged to supply these markets and have generated significant employment, should become part of the formal forest sector. Their access to forest resources should be limited to the nonpermanent (rural) domain. Sites should be chosen following consultation with and consent by neighboring rural people and local councils. Simple accountability mechanisms and straightforward taxation schemes could be developed gradually in collaboration with small-scale operators, and continued access to resources could be granted by communities based on performance.

These measures sound straightforward, but granting formal harvesting rights to small firms will be challenging. Given the chance to produce legally, artisanal loggers may find selling timber for the export market more profitable. Local markets would once again be undersupplied, and new illegal artisanal loggers would emerge to meet demand. Artisanal logging can be very hard to monitor outside well-established community forests or partnerships with larger companies. Distrust of bureaucratic processes and fear of harassment by forest and revenue department staff discourage artisanal loggers from operating legally.

In some instances, partnerships between small and large enterprises and innovative solutions to create a steady supply of material to undersupplied regions have started to emerge. They should be encouraged and replicated. Access to the more lucrative export market should continue to be subject to the law and to fair competition by all companies, large and small.

Give Greater Attention to Nontimber Forest Products

Nontimber forest products (such as medicinal plants, thatching, gum arabic, and bush meat) make a vital but still underappreciated contribution to the viability of forests and the national economy. These products are subject to uncontrolled and illegal exploitation, and arrangements to protect their sustained use must be considered in devising new arrangements to protect customary rights and encourage multiple, overlapping forest uses.

Attract New Eco-investors to Make Conservation Efforts More Sustainable

Cameroon is moving to a more holistic approach in forest management, based on multiple uses of forests and integrated production of forest products and global environmental services. To this end, Cameroon seeks partners to transform about 900,000 hectares of forest, originally destined for logging, into conservation concessions that will store carbon, harbor biodiversity, and yield other global goods. Arrangements for conservation concessions would need to include compensation for the economic and fiscal revenues forgone by removing the land from industrial use. Attracting qualified investors who can negotiate fair arrangements, especially with respect to enhancing local livelihoods, will be fundamental for Cameroon to retain a positive view of forest management options that are not based exclusively on timber.

Reshape Community Forestry

Even though the regulatory framework is being revised, major adjustments need to be made, starting at the conceptual level, if community forests in Cameroon are to represent a better option than business-as-usual land and tree management.

As a first step, the prospects of community forestry can be vastly and rapidly improved by removing the well-known impediments of costly inventories and excessive regulation (including environmental impact assessments, whose usefulness has yet to be demonstrated) and by permitting greater freedom in identifying the particular technologies and partnerships employed to manage these forests. The primary actions required are to adjust impeding regulations and issue and disseminate the new community forestry manual. This manual is the result of long and difficult discussions among several parties and represents a very significant improvement over the previous manual.

Simplified procedures will be more successful if communities learn how to implement them. It is imperative to strengthen community capacity to prevent other actors from assuming a disproportionate role in decisions related to community forests.

As a second step and more broadly, there is scope to rethink the entire community forestry approach and conceive community initiatives that can be implemented throughout the forest space whenever they are compatible with other forest uses (for example, in managed forest concessions and natural parks, where communities can partner with government or NGOs).

Community forestry support would then feature new schemes that would accommodate the products of a specific location, the opportunities presented by niche markets, and the profile of the rural investor, be it a community, an enterprise, or a cooperative. This may imply granting priority or exclusive rights to commercial use of specific timber and nontimber forest products to local enterprises, helping them with marketing information and legal advice, and offering other types of support.

A 2006 review of community forests (Cuny and others) highlighted the wide disconnect between the spirit of the 1994 Forest Law and the situation in the forests themselves, where the aims of reducing poverty and responsibly managing forests were far from universally respected. Based on the review's conclusions, a large consultation was organized with all parties involved (representatives of communities, the private sector, NGOs, and local and central administration), leading to a consensual proposal that supports changes in the conceptual framework for community forestry and designs a new regulatory framework.

Adapt Reform Instruments to Emerging Forest Management and Industry Needs

By 2007, 48 forest management plans had been approved and were being implemented by industrial forest operations in UFAs. Emerging evidence suggests that in Camcroon's mostly secondary forests the implementation of these management plans imposes severe constraints on species mix and volumes harvested. The new social, fiscal, and other obligations that often accompany these plans can make it more difficult for industry to integrate them into their business model.

The fragility of Cameroon's industry as it transitions to a highly regulated and managed forest production system is underappreciated because international timber prices have remained relatively high. High prices have somewhat compensated for the industry's limitations. In fact, industry continues to rely on a few tree species, remains focused on sawn timber, and has failed to invest in market research and technology that make it possible to use a broader range of species. Unless adequate measures are taken, several companies could be bankrupted by the next downturn in timber prices. Anticipating these risks, a series of measures has been proposed in a recent audit of Cameroon's forest sector (CIRAD 2006). These measures are presented in table 6.1; additional technical adaptations to better integrate the tax regime and auction system are discussed in appendix 10.

Table 6.1. Suggested adjustments to reform instruments to fit the new context

Finding	Suggestion
Mitigate risks associated with a fixed area tax	
The heterogeneity of forests in frequency and distribution of species, timber quality, and unproductive areas is not always considered in forest inventories (both sounding and large scale). An information asymmetry is always present, given limited public information about the condition of UFAs, and can lead to overbidding.	Limit the area tax to the concession's productive area, once the management plan has been approved (this measure also provides an incentive to complete plans expeditiously). After the auction, if the management plan unexpectedly reduces the potential yield (by increasing the minimum diameter required to harvest the main species), a corresponding reduction of the area tax can easily be calculated and enforced.
Given fluctuations in international timber prices, it is unlikely that firms can correctly reveal their "willingness to pay" based on expectations of future economic returns. Having to pay a fixed yearly area tax, when a large part of a firm's cash flow is determined by international prices, exposes the firm to risk when prices fall.	The area tax could be linked to the international price of tropical wood by identifying a basket of forest products (logs, sawn wood, ply and sliced veneer, and plywood) from different species, on the basis of which a wood price index would be calculated and regularly updated.[a]
Restore the financial capacity of forest companies and encourage investment	
The 2006 audit highlights the very low level of investment in new technologies in logging and processing units; (between 2001 and 2006) higher fixed costs (such as energy and transportation costs) and the tax burden have greatly eroded the financial capacity of most companies. Under the guarantee system, funds deposited by industry are designed to ensure financial compensation when taxes are unpaid or environmental regulations are violated, but this guarantee is never applied. The guarantee creates an unnecessary financial burden for companies and reduces their ability to access credit and invest.	Substantially reduce the bank guarantee for companies that have proven compliance with fiscal and environmental commitments during the last two years. Additionally, the Finance Administration should take strong measures to avoid delays in reimbursing the guarantee at the end of each year. Encourage certification (see below).
Promote diversification and use of secondary species	
Contrary to expectations, forest tax and management reforms did not expand the range of harvested species. Industry remains focused on a very few species. The tax burden seems to be one of the major constraints on diversification.	Significantly reduce taxes on less-commercial species. At the same time, enact more stringent controls, including at the mill gate, to prevent illegal activity.

Finding	Suggestion
Encourage certification	
Some companies operating in Cameroon have already been certified as practicing sustainable forest management, but they remain in the minority. Certification carries significant additional costs and risks (arising from third-party review and public disclosure of information). Given the advantages of certification both nationally and globally, fiscal incentives should be provided for companies to pursue certification.	Grant a reduced area tax to firms that exceed local requirements and obtain independent certification of their forest concession. Government must decide which certification systems to endorse and the duration of tax rebates for certified firms. This measure would strongly increase certification, which will accelerate the implementation of management plans (a condition for certification).
Add flexibility to accessing forest resources	
The Forest Law permits each annual harvestable area to remain open for three years. This practice allows operators to return to harvest species that had no market at the time of the initial harvest, and it allows them to delay timber sales when prices are low.	Permit operators to open an annual harvestable area one or two years in advance to enable them to take advantage of concessions when timber prices are high. This added flexibility must be accompanied by measures to preserve the integrity of rules for the harvesting cycle (under which a parcel must be closed for the duration of the cycle).
Suppress administrative dysfunctionalities	
Dysfunctional administrative services increase costs for industry in the form of informal taxes and fees, abusive financial cautions, excessive delays in refunding value added taxes paid by exporters, and so forth.	Improve governance within the administrations (specifically Forests and Finance) to make them more accountable to users. Promptly sanction any abuse of power by a civil servant. Enhance transparency through such actions as making administrative documents freely available online.

Source: CIRAD 2006.

a. Given that operators might be reluctant to disclose their prices to competitors, it is important to work with a price index. The International Tropical Timber Organization provides such information twice a month for a dozen African species (logs and sawn wood) and could be used as a starting point.

Final Thoughts: Can the Reforms Achieve Lasting Results?

Ten years of forest sector reforms have not been easy or straightforward. Much has been learned, and much remains to be done, but it is clear that some real achievements have been made.

A relatively well-conserved forest resource.   With more than 60 percent of its forests under biodiversity conservation, production forest management plans, and community or local council forests, Cameroon still has an economically and environmentally significant forest estate. The most serious threats against forest lands have virtually ended in permanent production forests. Deforestation remains contained to 0.14 percent per year according to most recent data (de Wasseige and others 2009) and occurs mostly outside permanent forest areas, concentrated in areas where communities have the freedom to convert forest land to agriculture or other uses. Increasingly, the greatest drivers of deforestation (agricultural expansion and domestic energy needs of urban centers) and solutions for the problem (agricultural intensification, commodity prices, land tenure security, and land markets) reside outside the forest sector.

Improved management practices.   The proportion of Cameroon's production forests covered by forest management plans (66 percent) is one of the highest among tropical countries. The quality of these management plans is considered better than those in Southeast Asia and Brazil (where the 30-year horizon is not used). Certification is progressing rapidly. Cameroon is generally ahead of other African countries with respect to transparency in allocating harvesting rights, public information, social redistribution of forest revenues, use of forests as local development assets, and devolution of responsibility to village communities.

Customary rights and social welfare.   Cameroon has advanced rules to preserve customary rights in all forests and is developing provisions to recognize and protect the rights of all people who depend on forests, Bantu as well as indigenous people. Despite its mixed record, community forestry appears to be capable of learning from its own mistakes and advancing further.

A restructured forest industry.   A significant part of industry has restructured, adapting its business model to cost structures that include higher fixed costs, new investments for management plans, and increased social and environmental responsibility. Overcapacity is less acute.

Formal, institutionalized collaboration to improve forest governance and transparency. Formal and institutionalized collaboration with NGOs has improved governance and transparency (independent observers for the auction process and forest control operations and a public register of forest and wildlife infractions are examples).

Can progress continue? Cameroon's innovative forest reforms are being observed with interest by other countries, especially Congo Basin countries, because Cameroon in many ways is the crucible in which potential regional approaches to forest reform are being tested. It is premature to speak of "final outcomes," yet the progress observed in Cameroon's rainforests is too significant to be overlooked. Will it continue?

It is certain that the forest sector reforms have laid a foundation that is not likely to vanish overnight—and may ultimately find expression in settings other than Cameroon. Several elements of the reforms seem relatively durable, such as the recognition that, like the public treasury, national forests are public property and should be treated with comparable care and subject to comparable accountability systems and controls. Similarly, government and industry appear to have accepted that forests are a finite entity, that long-term access to this resource is a valuable asset, and that the environmental and financial sustainability of this resource depends on sustainable management and industry efficiency. They also appear to recognize that third-party verification and disclosure have considerable power to ensure accountability. Civil society and parliament now expect to be involved in decisions pertaining to forests, just as communities expect to be involved in such decisions and to benefit from forests under community and industry management. With many forest staff retiring and significant resources becoming available from government and the donor community, government can go a long way toward sustaining the reforms by appropriating sufficient salaries and complementary funds to hire younger, better-trained personnel who can perform the more sophisticated regulatory and supervisory functions that are needed.

Other elements of reform appear more vulnerable. Community forestry reforms are recent, weak, imperfect, burdened by unrealistic requirements, and often the victim of outside interests. The prospects for sustaining responsible management of protected areas will improve substantially if donors and eco-investors respond positively to offers such as the establishment of the Ngoïla-Mintom conservation concession. Such opportunities are unlikely to exist indefinitely. Another vulnerability lies in the weak collaboration between the Ministries of Forests and Finance in collecting forest revenue. Finally, the competitive allocation of forest harvesting rights could be threatened. All large and more visible concessions will soon be allocated, and administrators may be uninterested in maintaining the same level of rigor in the auctioning of smaller harvesting rights or of concessions that are cancelled and/or resubmitted to public auction.

All of these risks must be followed closely, and opportunities to do so have been integrated into ongoing collaboration between the government and the community of development partners, such as the FESP and Forest Law, Environment, Governance, and Trade Initiatives.

Forest sector reform has engendered many very positive expectations among Cameroon's people and the international community. If Cameroon succeeds in consolidating the achievements of the past decade and addressing the unfinished reform agenda, it will substantially fulfill those expectations and improve economic, social, and environmental welfare, providing valuable lessons for other countries. The experience in Cameroon has shown that by widening and deepening the sources of information, and by providing information as a public good, a committed government can heighten awareness of institutional behavior and maintain the momentum toward greater accountability.

Perhaps the most urgent question arising from Cameroon's experience with forest reform remains to be answered. Can reform in one sector change the wider trajectory of society in any meaningful way? Or will the forest sector remain at variance with the rest of the country until it succumbs to the inertia that prevails elsewhere? It is hoped that forest sector reform will rebound with renewed support—from government as well as civil society—and will strengthen broader efforts to promote good governance and stewardship throughout the country.

Note

1. This committee, under the authority of the minister of finance, was the main interlocutor of the Bretton Woods institutions on all issues related to the SAC III reform and the Heavily Indebted Poor Countries process.

Appendixes

Structural Adjustment Credits, 1994–98

The forest sector became one of the focus sectors in the successive structural adjustment programs negotiated between the Bretton Woods institutions and Cameroon until the mid-1990s: the Economic Recovery Credit of 1994, and the second and third Structural Adjustment Credits (SAC) II and III, approved in 1996 and 1998, respectively. In the third SAC, the forestry sector became a full component of the credit.

Structural adjustment programs between 1994 and 1998. The forest component of the credit approved just after the January 1994 devaluation of the CFAF aimed at reducing sector expenditures by institutional consolidation, increasing government revenues, and supporting preparation of a new forest law. The World Bank was disappointed with the outcome of this program, mostly because of the features of the proposed forest law regarding concession award procedures, the log export ban, and an expanded role for government institutions. The approval of a satisfactory forest law therefore became a condition of effectiveness for an Economic Recovery Credit, approved in June 1994. A new adjustment program, SAC II, negotiated in 1996, was to help improve public finance management and in this context included forest taxation reform.

These conditionality-based operations, which enjoyed little government ownership, were conceived and implemented contentiously rather than collaboratively between the government and the World Bank. They resulted in paper reforms and contributed little to improving forest governance and sustainability on the ground.

The Third Structural Adjustment Credit (1998). This $150 million credit was intended to increase the country's short-term liquidity and establish reforms that would benefit the economy in the longer term. It included a fixed tranche common to four sectors plus a floating tranche. Common tranche disbursement was contingent on all sectors meeting specific conditions, a mechanism that generated pressure on ministries that lagged behind. Floating tranches were sector specific, and the nondelivery of one sector would not undermine disbursement of other sectors' tranches. Given the poor previous results and the risk that forest conditions would get in the way of disbursing SAC resources, both Bank staff and government officials hesitated to consider forests as a sector for SAC III.

In contrast to previous structural adjustments programs, SAC III contained several detailed and specific conditions related to forestry (box A1.1). Some were required for disbursement of the fixed tranche of the loan, while others were included in a floating tranche whose disbursement schedule was left open and was contingent on forestry conditionalities. This structure made credit disbursements more dependent on forestry reforms and created incentives for the government to accelerate the reform process, without tying all the reforms to a fixed schedule.

The forest component of SAC III aimed at creating regulations and institutional capacity to implement the 1994 Forest Law, which, in 1998 existed only on paper. The themes underlying its measures were to generate and test political commitment to the 1994 Forest Law, create the regulatory infrastructure to implement the Forest Law, design and enforce a new forest taxation system, and improve sector transparency, governance, and combat corruption. The reform package focused on fundamental policy issues that were essential to resolve to establish minimum sector governance conditions and eliminate distortions.

Since 1998, forest sector reforms have been woven into most if not all relevant development efforts that the World Bank conducted in coordination or collaboration with the IMF, other donors, and NGOs with a strong presence in Cameroon. The progress toward the completion point of the HIPC Initiative was also triggered by forestry measures. The most important of these was the respect of the SAC III reforms; the HIPC document clearly stated that the "satisfactory implementation of the SAC III reforms is a benchmark to the achievement of the HIPC completion point." More specifically, special attention was given to enforcing the for-

Box A1.1

Summary of Third Structural Adjustment Credit forest measures

Measures to be met prior to submission of the Credit for the approval of the WB Board of Executive Directors

- Cancellation of concessions awarded in October–November 1997, for which agreements have not been signed or guarantees/deposits paid.
- Demonstration of the government of Cameroon's willingness to review jointly with the World Bank the 1997 allocation round and grant new forest management units (UFAs) and short-term harvesting rights (ventes de coupe) based on the recommendations of the joint review.

Measures for effectiveness

- Adoption of a new, transparent, and competitive system, including the services of an independent observer, for the allocation of forest harvesting rights.
- Definition of a forest concession planning strategy, taking into account requirements for sustainable forest management.
- Creation of a forestry tax revenue-enhancement program, to improve monitoring and recovery of forest fees and taxes.

Measures for release of floating tranche

- Adoption of procedures for the preparation, approval, and monitoring of forest management plans.
- Establishment of a guarantee system to ensure compliance with forest management plans.
- Selection of an international consultant to assess the implementation of forest management plans on the ground.
- Completion of an independent economic and technical audit to assess the sector's economic and fiscal revenue potential.
- Submission to parliament of reforms concerning forest industry development and the taxation regime.
- Selection of areas to be reserved as community forests.
- Revision of the National Forestry Development Agency's institutional status, clarifying its mandate and sources of financing.

estry laws and regulations in the field and the prosecution of contentious cases.

Once forest sector reforms had demonstrated their relevance to the success of broader national reforms (such as public finance, procurement, fiscal responsibility, and law enforcement), it was only natural for these to be featured and monitored in the context of the Letters of Intention to the IMF, the Poverty Reduction and Growth Facility, and the HIPC Initiative.

Similarly, given the role of forest legislation and institutions for the success of environmental mitigation plans and the credibility of environmental offsets, forest reforms and related issues have been mentioned and have featured in most prominent World Bank–supported interventions, such as the Chad-Cameroon Pipeline and the Railway Rehabilitation Project. The convergence of interests and political imperatives that emerged between the forest agenda and major policy macroeconomic frameworks and development projects further elevated the political standing of forest reforms and created stronger incentives for those interested in its success.

Production Forests Gazetted as of February 2006

FMU number	Concession number	Name	Provisory area (ha)	Definite area (ha)	Difference (ha)	Percent	Status
FMUs allocated in 1996							
7							
10.001	1025	CFC 1996	63,728	69,018	5,290	8	Gazetted
10.002			28,026	22,784	−5,242	−23	Gazetted
10.003			67,217	48,830	−18,387	−38	Gazetted
10.004			44,651	52,473	7,822	15	Gazetted
08.001	1026	EGTF RC CORON	61,760				Department meetings
08.002			75,000	59,910	−15,090	−25	Gazetted
07.002	1027	CPPC Ex CELLUCAM	100,000	100,000	0	0	Gazetted
FMUs allocated in 1997							
19							
08.003	1020	SFH/STJJY Sarl	53,160	45,210	−7,950	−18	Gazetted
08.004	1017	EFMK	126,160	88,050	−38,110	−43	Gazetted
08.006	1002	SFB	69,920	51,450	−18,470	−36	Gazetted
09.006	1001	SFF	74,092				District meetings
09.021	1006	WIJMA	41,965	41,965	0	0	Gazetted
09.023	1005	BUBINGA	56,192				Department meetings
09.025	1011	SCIEB	86,788	88,148	1,360	2	Gazetted
10.007	1010	SEBC	113,507	122,294	8,787	7	Gazetted
10.009	1022	SEBAC	88,796	92,287	3,491	4	Gazetted
10.011	1013	SAB	60,838	48,554	−12,284	−25	Gazetted
10.012	1016	SEFAC	62,597	59,340	−3,257	−5	Gazetted

FMU number	Concession number	Name	Provisory area (ha)	Definite area (ha)	Difference (ha)	Percent	Status
10.018	1003	SIBAF	65,832	81,397	15,565	19	Gazetted
10.021	1018	Green Valley	71,533	66,183	−5,350	−8	Gazetted
10.023	1007	SFCS	62,389	57,996	−4,393	−8	Gazetted
10.029	1014	SFBD	46,922	46,922	0	0	Gazetted
10.041	1019	PALLISCO	64,961	64,961	0	0	Gazetted
10.051	1015	GRUMCAM	85,812	86,096	284	0	Gazetted
10.054	1012	SFID	68,292	67,942	−350	−1	Gazetted
10.058	1009	SEBC	60,823	57,137	−3,686	−6	Gazetted
FMUs allocated in 2000							
23 00.003	1028	MMG	125,568	125,568	0	0	Gazetted
00.004	1029	TRC	125,490				Inactive since 2000
08.008	1030	SCTCB	72,000				Department meetings
08.009	1031	INC	65,472	49,640	−15,832	−32	Gazetted
09.004b	1033	COFA	81,335				District meetings
09.003, 09.004a, 09.005a	1032	LOREMA	138,652	141,273	2,621	2	Gazetted
09.005b	1034	SOCIB	40,297	43,010	2,713	6	
09.015	1035	SN COCAM	41,559				District meetings
09.019	1036	CUF	38,247				Department meetings
09.024	1037	WIJMA	76,002				Department meetings

FMU number	Concession number	Name	Provisory area (ha)	Definite area (ha)	Difference (ha)	Percent	Status
10.020	1038	ING FOR	87,192	82,571	−4,621	−6	Gazetted
10.022	1039	SCIFO	48,864	35,090	−13,774	−39	Gazetted
10.026	1040	ALPICAM	128,449	126,988	−1,461	−1	Gazetted
10.031	1041	ING FOR	41,202				District meetings
10.037	1042	KIEFFER	51,685				District meetings
10.038	1043	CAMBOIS	145,585				District meetings
10.039	1044	ASSENE NKOU	47,585				District meetings
10.045	1045	J. PRENANT	54,447				District meetings
10.046	1046	SCTB	70,283				Department meetings
10.062	1047	PANGIOTIS	138,675	149,079	10,404	7	Gazetted
10.063	1048	SIBAF	68,933	68,916	−17	0	Gazetted
FMUs allocated in 2001							
16 08.007	1049	SABM	83,400				Department meetings
09.017, 09.018	1050	FIPCAM	99,501				District meetings
10.005a	1051	STBK	52,245	89,322	0	0	Gazetted
10.005b		nonallocated	37,077				
10.008	1052	SEFAC	60,053	72,727	12,674	17	Gazetted
10.010	1053	SEFAC	61,670	66,688	5,018	8	Gazetted
10.015	1004	CIBC	155,421	130,273	−25,148	−19	Gazetted
10.030	1054	PALLISCO	79,757				District meetings

FMU number	Concession number	Name	Provisory area (ha)	Definite area (ha)	Difference (ha)	Percent	Status
10.042	1055	SODETRANCAM	44,249				District meetings
10.044	1056	ASSENE NKOU	66,861				District meetings
10.047a	1057	FIP-CAM	47,381	47,080	–301	–1	Gazetted
10.052	1058	SOTREF	69,008				Department meetings
10.059, 10.060	1059	SCTB	93,174				Department meetings
10.061	1021	PLACAM	27,495	28,387	892	3	Gazetted
10.064	1060	FILIERE BOIS	114,379	115,900	1,521	1	Gazetted
FMUs allocated in 2002							
7 09.012	1062	AVEICO	85,131				District meetings
09.013	1063	CFK	50,296				District meetings
09.016	1064	COFA	66,007				District meetings
10.013	1065	CFE	70,441	50,752	–19,689	–39	Gazetted
10.056	1066	SFID	72,391				Department meetings
10.057	1067	ING FOR	32,293				Department meetings
11.002	1068	WIJMA	72,705				Transmitted to PM
FMUs allocated in 2005							
14 00.001, 00.002	1075	SEPFCO	73,336				Department meetings
08.005	1077	SIM	56,900	36,340	–20,560	–57	Gazetted
09.007, 09.008	1073	MPACKO	79,422			.	
09.009, 09.010	1076	SFB	81,835				Department meetings

FMU number	Concession number	Name	Provisory area (ha)	Definite area (ha)	Difference (ha)	Percent	Status
09.020	1069	CUF	44,866				Department meetings
09.022	1078	GAU Services	78,461				Department meetings
10.025	1070	SFIL	76,002	47,823	−28,179	−59	Gazetted
10.040	1074	TTS	79,579				Department meetings
10.043, 10.055	1071	PLACAM	92,241				Department meetings
10.053	1072	GRUMCAM	82,308				Department meetings
FMUs allocated in 2006							
10 09.011		SIBM	35,891				Department meetings
09.014		GEC	28,931				
09.026, 09.027		CUF	64,461				Department meetings
09.028		Effa JB & Cie	26,895				—
10.047b		CCIF	48,960	48,960	0	0	Gazetted
10.048		SCIFO	68,030				Department meetings
10.049, 10.050		PFM Wood	70,688				Department meetings
11.001		TRC	55,580				Department meetings

FMU number	Concession number	Name	Provisory area (ha)	Definite area (ha)	Difference (ha)	Percent	Status
Nonallocated FMUs—Conservation concessions							
09.001			180,606				—
09.002			76,840				—
10.027			72,465				—
10.028			91,063				—
10.032			97,242				—
10.033			48,754				—
10.034			163,952				—
10.035			99,515				—
10.036			65,055				—
TOTAL Conservation concessions			**895,492**				

Source: Public records of the Ministry of Forests and Wildlife, Cameroon

Note: FMU = forest management unit

Conservation Estates in Cameroon

Type	Name (type)	Area (ha)
National parks	**Benoué**	**180,000**
	Bouba Ndjida	**220,000**
	Boumba Bek	**238,255**
	Campo Ma'an	**264,064**
	Ebo	*143,000*
	Faro	**330,000**
	Kalamaloué	**4,500**
	Korup	**125,900**
	Lobéké	**217,854**
	Marine Park of Kribi	—
	Mbam et Djérem	**416,512**
	Mefou	*1,044*
	Mont Bakossi	*36,000*
	Mont Kupé	*4,300*
	Mozogo Gokoro	**1,400**
	Mpem et Jim	**100,000**
	Ndongoré	*230,000*
	Nki	**309,362**
	Takamanda	*69,599*
	Vallée du Mbéré	**77,760**
	Waza	**170,000**

Type	Name (type)	Area (ha)
Wildlife sanctuaries	**Banyang Mbo Gorilla Sanctuary**	**66,000**
	Kagwene Sanctuary	*1,100*
	Lom Panghar Gorilla Sanctuary	*47,686*
	Mengame Gorilla Sanctuary	*96,050*
	Mont Oku	*20,000*
	Mont Oku Sanctuary	**1,000**
	Rumpi Hills Sanctuary	*45,675*
Wildlife reserves	**Cratère de Mbi**	**370**
	Dja	**526,000**
	Douala–Edéa	**160,000**
	Kimbi	**5,625**
	Kupé (integral ecological reserve)	*4,676*
	Lac Ossa	**4,000**
	Manenguba (integral ecological reserve)	*5,252*
	Mont Bamboutos	*2,500*
	Santchou	**7,000**
Zoological gardens	**Garoua**	**2**
	Limbé	**1**
	Mvog Betsi	**2**
Total area of protected areas	*Areas in the process of being created and gazetted*	*706,882*
	Areas officially created and gazetted	**3,161,543**

Source: Authors.

Note: Names in **bold** indicate areas officially created and gazetted. Names in *italics* indicate areas in the process of being created and gazetted.

Bidding System
for Harvesting Rights

The bidding process involves two stages, the first of which is the technical evaluation. Financial offers are considered only from bidders that meet or exceed the minimum technical requirements. The formula for determining the winning bid is (technical score × 0.3) + (financial score × 0.7).

There has been sharp debate on the effects of the 70:30 weighting ratio given to financial and technical scores. Two arguments favor assigning a heavy weight to the financial offer. First, adequate thresholds in the technical qualification criteria ensure that all operators participating in the auction are fully technically qualified. Second, the comparison of financial values is simple and straightforward, offering the best guarantees of objectivity.

Would a weighting ratio that gives more importance to technical criteria result in lower financial offers? It probably would not, because bidders would still try to maximize their chances of success by making higher financial offers.

In that case, would the weighting system favor wealthier and non-Cameroonian firms? Yes. There is a trade-off between efficiency and wider access to the resource. Even so, a number of domestic operators have had winning bids in the auctions, and some UFAs are earmarked for bidding by Cameroonian firms only. Medium-size domestic firms also appear to have formed more partnerships with larger foreign firms to deal with the higher costs and meet the requirements for participating in the auctions.

Small-scale domestic interests have the alternative of pursuing harvesting rights through community and local council forests.

Would assigning more weight to technical capacity improve the selection of "good" operators and improve forest management? The evaluation of the financial offer is objective by nature, but the evaluation of technical capacity (which includes the investments that firms promise to make to ensure good forest management) is obviously more subjective and has the potential to introduce vested interests into the selection process. The independent observer seems to have acknowledged this risk in stating that a "stable doctrine" (in other words, a clear set of criteria) was needed to assess technical capacity. Clearly the evaluation of technical capacity can result in contradictory and unreliable outcomes, even if technical capacity is evaluated without interference. In the absence of a body to monitor whether concession holders fulfill their commitments to invest in technical capacity, sanctions for noncompliance (for example, fines or the loss of the concession), and an administrative commission to assess technical capacity (rather than the competing firm itself), it would appear that assigning more weight to technical capacity would simply increase corruption and abuse.

Chronology of Changes in the Legislative Framework for Forest Taxation

Changes in the legislative framework for forest taxation

The 1994–95 finance law
- Increase of RFA (annual area tax) from CFAF 98 to CFAF 300/ha/year.
- Establishment of export duties on processed export products.

The 1997–98 finance law
- Increase of RFA floor price to CFAF 1,500/ha/year for UFAs (forest management units) and CFAF 2,500/ha/year for a sale of standing volume. (It is worth mentioning that this increase caused a general outcry from the industry.)
- Change of the basis of calculation of timber value from a price-list system not in line with the international market to fob values provided by SGS-Forestry and the Reuters Agency.
- Increase of the sawmill entry fee rate to 12.5 percent for factories under common law and 17.5 percent for special industrial free zone factories.
- Institution of a 17.5 percent tax on log exports.
- Institution of an auction system for allocating UFAs and ventes de coupe (VCs).

The September 1997 agreement between the government and logging operators

- Fifteen percent reduction of the fob values provided by SGS-Forestry and Reuters that were used as the tax base for the felling tax and export duties. This reduction was motivated by the fact that SGS and Reuters fob values reflected superior quality (Loyal Merchant) timber.
- Reduction of the sawmill entry fee to 3 percent for factories under common law and 4 percent for special industrial free zone factories.

Partial prohibition of log exports during the 1999 financial year

- Export of traditional species (sapelli, sipo, iroko, and others) was prohibited while export of species to be promoted and other lower commercial value species (ayous and azobé) was maintained.
- The approval of an economic and financial audit of the forestry sector.

The 2000–01 finance law (completed by the Decree 2001/1034/PM)

- The floor price for UFAs was reduced from CFAF 1,500/ha/year to CFAF 1,000/ha/year; the floor price for VCs was maintained at CFAF 2,500/ha/year.
- A bank deposit guarantee of all fiscal and environmental obligations, equal to the amount of the RFA, became mandatory. As a consequence, 3 of the 21 successful bidders for the July 2000 concession auction lost their concession after failing to pay the bank guarantee.
- Creation of a forest revenue redistribution system for local councils and local communities (respectively 40 percent and 10 percent of RFA revenues).
- A sawmill entry fee of 2.25 percent of the fob value of timber entering the sawmill was established.
- A 17.5 percent admission fee for timber entering sawmills located in the industrial free zone was instituted, causing most companies to abandon the free zone regime.
- Export duties on processed timber were cancelled in compensation for the increase of the area and the sawmill admission fees.
- Changes in the rates of the surtax on exported logs introduced as follows: CFAF 4,000/m^3 for ayous, CFAF 3,000/m^3 for first category promotion species (like azobé), and CFAF 500/m^3 for second category promotion species.

- The option of setting up a competitive export quota system is still under consideration (the implementation of this measure is still being analyzed).
- The option of creating a compensation or equalization fund to improve the apportioning of resources—a *fonds de péréquation*— is still on the table. Regulatory texts to specify the functioning of such a fund are still expected.

The 2003 finance law

- A provision allowing presumptive sanctioning of companies cited for violating forestry laws. A company that wishes to challenge a citation must pay half of the scheduled fine and provide a bank guarantee for the other half before going to court. It recovers these amounts only if the court dismisses the charges. An Instruction by the Director of Taxes (dated February 6, 2003) specifies the modalities of the presumptive sanctioning.
- The Finance Law defines the modalities for the logging activities during the transition period (July to December 2003).

The 2004 finance law

- Creation of the "Division des Grandes Entreprises" within the Directorate of Taxes of the Ministry of Finance, in charge of recovering all taxes (sector-specific and general) from firms with an annual turnover exceeding CFAF 1 billion.
- Activities generating profits within community forests are subject to common taxation (art. 9).

The 2005 finance law

- Export authorizations delivered upon the presentation of a release for forest taxes payment (including area fee and felling and sawmill entry taxes). In case of failure to pay these taxes, a 400 percent increase is applied, and exporters and producers are jointly liable.

2005 Forest and Environment Sector Policy Letter

This appendix presents the English translation of an original in French signed by the Government of Cameroon.

1. ACHIEVEMENTS

1. Sector Resources and Policy

The dense humid forests of Cameroon cover a surface area of about 19.598 million hectares (that is 41.2 percent of the national territory) in the southern part of the country, which also has forested savannahs and forest galleries covering about 4.3 million hectares. According to the overall assessment of surface area carried out within the framework of the National Environment Management Plan of 1996, dense forests and wooded savannahs cover a surface area of 21.070 million hectares, some 45 percent of the national territory. However, these forests have undergone and continue to undergo significant changes as a result of logging, exploitation of natural resources, and transformation into farmlands and for other uses.

In order to ensure the growth and sustainable development of the forest sector, the Government of Cameroon has over the last ten years initiated far-reaching reforms touching on the institutional and regulatory frameworks of the sector. Such reforms were embodied in the creation of the Ministry of Environment and Forests in 1992, the formulation of a new Forestry Policy published in 1993, the adoption of a new Forestry

Code in 1994 and an Environment Code in 1996. The Ministry of Environment and Forests gave way to the two successor Ministries established by Decree number 2004/320 on 8 December 2004: the Ministry of Forests and Fauna (MINFOF) and the Ministry of Environment and Protection of Nature.

The new forestry policy is an important component of the National Environment Management Plan. It also falls in line with the agricultural policy. The 1994 Law to lay down Forestry, Wildlife and Fisheries Regulations and the 1996 Law to institute the Framework Law relating to the management of the Environment, establish a policy and strategy framework that centers on the following aspects:

- Sustainable management of forests, with the creation of a permanent forest estate and the setting up of Forest Management Units (FMUs) to replace forest permits.
- Contribution to economic growth and poverty alleviation by ceding part of the income from tax revenue to councils, creating jobs and allocating community forests.
- Participatory management through consultation with the civil society and the private sector, sensitisation of rural populations about their responsibilities, and permanent dialogue with the international community.
- Conservation of biodiversity through a national network of protected areas.
- Building the capacity of the public sector in the performance of its traditional key functions and the transfer of productive functions to the private sector.
- Putting in place a legal framework conducive to the development of the private sector, based on long-term conventions and industrialisation.
- Harmonisation of the regional management system through a zoning plan.
- Improvement of governance through increased transparency and systematic dissemination of information to the public.

Besides, this policy is given greater momentum by government guidelines that lay emphasis on new policy challenges such as poverty alleviation, decentralization and good governance. It is also consistent with international guidelines on the environment laid out at the Rio Earth Summit in 1992 and Johannesburg Summit of 2002. An Emergency Action Plan (EAP) supplemented these actions in 2000, by laying emphasis especially on the fight against poaching and overall streamlining of the sector. The Emergency Action Plan itself has been integrated into the Forest and Environment Sector Programme (FESP).

2. Cameroon's commitments at regional and international levels

During the Central African Heads of State Summit in March 1999, Cameroon reaffirmed its national and sub-regional commitment to ensure sustainable management of the forest ecosystems of the Congo Basin. Concerted management of sub-regional forest resources decided upon by Heads of State in the Yaoundé Declaration of 17 March 1999 thus received the backing of the international community through resolution No. 54/214 of the United Nations. The Convergence Plan that resulted thereof and was approved by the Conference of Ministers in Charge of Forests in Central Africa (COMIFAC), was adopted at the last Heads of State Conference of the Central African Forestry Commission on 5 February 2005 in Brazzaville. This Plan is a compilation of the National Specific Programs of signatory States, which should contribute to the attainment of convergent objectives that translate into reality the resolutions of the Yaoundé Declaration of March 1999. Cameroon's FESP, which is the Convergence Plan implementation program at national level, will strive to be the national response to COMIFAC programs.

Cameroon's commitment to ensure sustainable management of natural resources at the international level has been translated in the signing of several regional and international conventions on forest and biodiversity. These conventions include the following:

- The Agreement to set up the Lake Chad Basin Commission (1964)
- The Convention on the Conservation of Nature and Natural Resources (Algiers, 1968)
- The Convention on the Protection of Cultural and Natural Heritage (Paris, November 1972)
- The Convention on International Trade in Endangered Species (CITES – Washington, March 1973)
- The African Timber Organisation (ATO – Bangui, 1974)
- The Agreement on the Joint Management of Flora in the Lake Chad Basin (Enugu, December 1977)
- The International Agreement on Tropical Timber (Vienna, 1983)
- The Central African Cooperation Agreement on Wildlife Conservation (Libreville, April 1983)

- The Vienna Convention on the Protection of the Ozone Layer (Vienna, March 1985)
- The Montreal Protocol on the Control of Chlorofluorocarbons (Montreal, September 1987)
- The Convention on Climate Change (June 1992)
- The Convention on Biodiversity (Paris, October 1994)

3. African Forest Law Enforcement and Governance (AFLEG)

Like most African countries, Cameroon committed to be part of the initiative launched in 1989 by the G-8 countries to combat illegal exploitation, illicit trade in forest products and corruption in the forest sector.

This initiative aims to stimulate, at the highest political level, the will of the international community to curb these scourges in order to ensure sustainable management of our forest resources. This explains why the efforts made by Cameroon to meet this imperative have, over the last ten years, been marked by political, legislative and institutional reforms, and adoption of basic instruments that should promote conservation and sustainable management of natural forests.

This concerns especially the following:

- Adoption of a forest exploitation system that makes use of development plans aimed at ensuring sustainable management of production forests;
- Improvement of forest governance by setting up a legal framework conducive to the development of the private sector, which guarantees transparency in the process of allocating forest concessions and supervising exploitation activities, the major innovation being the use of call for tenders and the presence of independent observers;
- Adoption of Principles, Criteria and Indicators (PCI) for sustainable management of forests in Cameroon in December 2004 is one of the links in the certification chain.

Progress made in terms of reforms geared towards the conservation and sustainable management of forests in Cameroon is remarkable, and thus deserves to be appraised and evaluated. However, this is an exacting process that requires the collaboration of all stakeholders in the sector.

Cameroon is determined to pursue these reforms. The concern here is to ensure rational management of a national heritage of great importance while mainstreaming future needs. We do not have the right to deprive future generations of the benefits that we are deriving from the forest today.

4. The Sector Reform Plan

On the basis of progress made and in order to consolidate the first phase of the essentially institutional and legislative reform of the forest sector, a second phase of reforms was launched with the support of the Bretton Woods Institutions, who mainstreamed a forest sector component into the Third Structural Adjustment Credit (SAC III) granted to Cameroon. This component is based on the following three key objectives of the forestry policy:

- Sustainable management of the resource;
- Stimulation of economic growth and contribution to poverty alleviation especially by ceding to councils and communities part of forestry revenue, job creation in the timber sector and community forests managed by communities themselves; and
- Development of a dynamic and efficient private sector.

The main reforms encouraged include the following:

Competitive allocation of exploitation permits. Forest exploitation permits (Forest Management Units, Sale of Standing Volume) are allocated by means of a competitive procedure on the basis of technical and financial criteria. File selection is carried out by an interministerial commission that includes an independent observer who contributes to improve transparency in the permit allocation procedure.

Planning of allocation of exploitation permits. In order to ensure rational management of its natural resources, the Government of Cameroon designed a concessions allocation planning strategy (UFAs) in 1999 and provided clarifications concerning the different types of existing permits. This strategy was updated in April 2004 and envisages the allocation of all forest concessions by 2006. Such planning constitutes an important instrument both for forestry and taxation services as well as forest exploitation companies.

Setting up the Forestry Revenue Enhancement Programme (PSRF). Placed under the twin supervision of the Ministries in charge of Forestry and Finance, the PSRF was set up in June 1999 to strengthen the hitherto very weak forestry-tax collection capacity. The PSRF centralises the collection of forest exploitation-specific taxes and common law taxes for big companies. Since the creation of the programme, the Ministry of Finance and Economy has recorded a significant increase in forest tax revenue (excluding export taxes): from CFAF 10 billion in 1998/1999, CFAF 29 billion in 2000/2001 to CFAF 40 billion in 2002/2003, at a time when the taxation system is fully operational and recovery rates reach 90 percent.

Forestry taxation reform. An economic and fiscal audit was conducted under the joint responsibility of the Ministries of Finance and Forestry by an internationally recognised Committee of Expert Economists. The objective of the audit was to redefine the framework for sharing the revenue deriving from the exploitation of forest resources between the Government of Cameroon, private operators and the population. The final report of that audit, which was validated in March 2000, made a number of recommendations which gave rise to major reforms adopted by the National Assembly. Such reforms include:

(i) Competitive allocation of forest exploitation permits;
(ii) Establishment of a deposit system;
(iii) Clarification of industrial free zones;
(iv) New forest products taxation modalities.

Some reforms are still not implemented. These include the quota system for timber exports and the setting up of a system for equitable distribution of the share of forest royalties paid to local councils and communities (Equalisation Fund).

Definition of the rules relating to forest development plans. Order No. 222/MINEF of 25 May 2001 defines the rules for the preparation, design and monitoring/control of development plans in forest concessions. It provides a detailed practical guide to all stakeholders involved in the sustainable management of forest resources.

Establishment of a bank deposit system. Forest exploitation permits are granted to operators after due payment of a bank deposit that covers all the risks of non-compliance with fiscal and environmental commitments. This system is aimed at discouraging speculative practices which are very rife in the forest sector.

Supervision of management plans. Partnership with two internationally recognised environmental NGOs has made it possible to strengthen the ability of the Ministry of Forestry to supervise management plans and allows for operational monitoring of exploitation activities in forested areas in Cameroon. The convention signed between the Government and Global Forest Watch—World Resource Institute (GFW), allows for the monitoring of forest exploitation activities in permanent and agro-forestal estates by means of advanced satellite imagery technologies. A map of the road network within FMUs and Sale of Standing volumes is produced each year. A second contract has been signed with

GW (Global Witness) which is expected to participate as independent observer in the supervisory activities carried out by the Ministry in charge of Forestry. Such activities range from field missions to the handling of all litigations. GW ensures that supervision procedures are upheld at every stage and is responsible for signalling any abnormalities to the Ministry of Forestry.

Promotion of community forests. Order No. 518/MINEF/CAB of 21 December 2001 offers local communities the possibility to use their right of pre-emption over surrounding forests. This right makes it possible for them to suspend any allocation of the forest concerned to a trading company for a period of two years during which the community may choose to transform it into a community forest.

Restructuring ONADEF. Such restructuring is in conformity with the objective of redistributing forest sector roles, with government services concentrating on traditional roles (policy formulation, supervision and control) and productive functions being delegated to private operators. Presidential Decrees No. 2002/155 and 2002/156 of 18 June 2002 modified the designation, the articles of association, the mandate and funding modalities of ONADEF (National Office of Forest Development). The restructuring of ONADEF gave birth to the National Forestry Development Agency (ANAFOR) whose mission is to promote economically profitable and environment-friendly forest plantations both in the private sector and in the local communities.

2. THE CHALLENGES TO OVERCOME

1. Harmonisation of institutional capacity and reform objectives

There is a big disparity in terms capacities for forestry and wildlife legislation implementation. Besides, there is no framework for the harmonisation and coordination necessary for the smooth development of various support projects. In addition, the support of the international community is very scattered and is translated by a great many uncoordinated projects, ill-appropriated by national institutions and with moderate impacts on the field. Over 60 projects under the supervision of the Ministries of Forest and the Environment and over 10 under the supervision of other Government services (mainly the Ministry of the Plan, Programming and Regional development, the Ministry of Agriculture and Rural Development and the Ministry of the Economy and Finance) currently have components that are more or less related to the protection, renewal and valorisation of natural resources. These projects annually take up close to CFAF 10 billion of external funding in the forest-environment sector alone, to which should be added the national counterpart funds.

2. Enhancement of popular participation in the management of forest and wildlife resources

In spite of the prescriptions of the current forestry policy that recommends the participation of the population in the conservation and management of forest resources, such participation remains limited. This is due especially to the unclear definition of the roles of the civil society and the local communities, which do not have adequate skills and organisational capacity needed for proper management of the redistributed resources, the weak technical management capacity of NGOs and the great distrust between them and the forestry administration, the difficulties for economic operators to move from the stage of "logger" to the stage of management forester.

3. Securing a significant share of forest revenues for community development

During the period from 2000 to 2004, over CFAF 25 billion was distributed to councils and local governments as their share of the royalty from the acreage under exploitation for commercial purposes. An audit and other studies were also carried out on the use of almost all of the said funds. In spite of these positive developments, much is still to be done to allow for more transparent and participatory planning of investments and more efficient control of expenditure.

4. Stepping up the fight against poaching and illegal exploitation of natural resources

The ability of the Ministry of Forestry and Wildlife to carry out controls, follow up offences and mete out sanctions is still relatively limited. In spite of the involvement of an independent observer in the control activities of the Ministry of Forestry and Wildlife, performances are still moderate. Only greater organisation of all the phases of the control chain by the Ministry of Forestry and Wildlife may ensure lasting results and make it possible to build on the contributions of the independent observer, especially by publishing the record of offences and informing the public about the litigations between the Ministry and forest sector economic operators.

5. Speeding up the mainstreaming of environment-friendly measures in forest production activities

Over six years after the entry into force of the Framework law on the management of the environment in Cameroon, environmental concerns have remained administrative formalities within the context of forest management operations. It is therefore necessary to ensure rigorous application of environmental management plans, especially in some forest management units situated near protected areas where environmental assessments are a key precondition. Besides, it will be necessary to strengthen and step up the operational abilities of the institution in charge of the environment and of advisory bodies such as the National Advisory Committee on the Environment and Sustainable Development (CNCEEDD).

6. Protected areas—National parks

A network of protected areas has been developed especially over the last fifteen years, thanks to the Biodiversity Conservation and Management Programme in Cameroon (PCGBC). Wildlife protected areas cover 7 211 819 hectares representing 15.1 percent of national territory.

This surface area is distributed as follows: 10 National parks; 10 Wildlife reserves; 1 Wildlife sanctuary; 3 Zoological gardens; 40 Hunting zones; 16 community-based hunting zones which have been placed at the disposal of village communities. The technical and financial management of community-based zones is the responsibility of the populations who may, if they so desire, seek the technical assistance of the staff of the Ministry of Forestry and Wildlife.

Besides, a method of assessing efficient management of all national parks has been established.

7. Expansion and participatory management of protected areas

The different measures taken to this effect include the following:

a) Economic operators in leased hunting zones are obliged to respect the specifications which stipulate the social projects to be carried out for the benefit of the surrounding population.

b) In addition to such social projects, communities benefit 50 percent of lease taxes which are annual and per hectare, and shared on a pro-rata basis of 40 percent to the councils and 10 percent to communities. Cameroon also instituted rules that allow for consultation of the local populations both at the level of creation, demarcation and management of protected areas.

c) Consolidation of protected areas within the national territory and setting up of cross-border protected areas. The Conventions relating to the Sangha Cameroon (Lobéké National Park), the C.A.R. (Dzanga

Sangha National Park) and Congo (Nouabale Ndoki National Park) complexes as well as the Gahanga-Gumti (Nigeria) and the Tchabal Mbabo (Cameroon) complexes have been signed. Agreements between the Dja Wildlife Reserve (Cameroon), Minkébé (Gabon), Odjala Reserve (Congo) on the one hand, and the Korup (Cameroon) and Cross River (Nigeria) on the other hand are being prepared.

d) Fight against poaching, especially through the setting up of a central Anti-Poaching Unit, a National Committee and Provincial Committees to this end. Such committees involve transport companies (CAMRAIL and CAMAIR).

These measures are coupled with regional initiatives such as CEFDHAC, COMIFAC, OCFSA, ECOFAC, WWF, CARPE and compliance with standards and international conventions (CITES).

8. Extension of programmes to arid and mountainous regions
As a result of its geographical position, Cameroon is part of the sudano-sahelian zone threatened by desertification. Forest ecosystems are fragile, often because of overexploitation by the population for subsistence.

Research carried out over the last ten years on species, modes of settlement and management, makes it possible to propose suitable solutions for reforestation, agro-forestry and management of rural and urban peripheral zones with a view to fixing the soil, restoring fertility, producing fuel wood and lumber, fodder, food products and financial revenue from the sale of forest products.

Urban periphery and agro-forestry reforestation programs in savannah and forested areas shall be partly implemented (directly or indirectly) by ANAFOR in accordance with the national program for the planting of private and community forests.

3. PROSPECTS AND COMMITMENTS FOR 2005–2015

Future prospects centre on solving problems and reviving the sector. They are based on the consolidation of the reform framework established through SAC III and the revival of sector development programmes through the Forest and Environment Sector Programme.

1. Consolidation of the reform framework
In spite of the progress that Cameroon has made during the last ten years towards sustainable management of forest resources and improvement

of governance and transparency, illegal exploitation of resources is still a disquieting phenomenon that jeopardises all of the progress already made. This phenomenon can only be entirely eradicated and forest resources sustainably managed if we commit to uphold the SAC III reform framework and pursue its implementation. The table below is a recap of the reforms introduced and highlights overall reform implementation indicators.

Reforms	Reform implementation
Competitive allocation of exploitation permits	Transparent allocation of forest exploitation permits (quality of bidding documents) Participation of an independent observer in the permit allocation committee meetings and forwarding of reports Exclusion of companies convicted of major infringements of forestry legislation and regulations
Planning of forest concessions allocation	Allocation of forest concessions according to the programme elaborated
Forestry Revenue Enhancement Programme	Collection of forest-sector specific taxes (PSRF annual report) Collection of fines and damages arising from offences (PSRF annual report) Effective payment of all bank deposits (PSRF annual report) Setting up a system for fair redistribution of RFA shares (equalisation fund)
Forestry taxation reform	Pursuance of fiscal reform implementation Effective and efficient consultation between the Government (MINEF-MINFB) and the private sector
Sustainable management of forest concessions	Regular meetings of the Technical assessment committee of management plans and the Inter-ministerial Committee for the approval of management plans Number of concessions allocated since two years that have management plans approved by the administration Number of final conventions signed
Supervision of management plans	Regular monitoring of forest exploitation with satellite imagery (road map available each year) Participation of an independent observer in MINEF control activities
Promotion of community forests	Transparent application of the right of pre-emption for local communities Effective start of the RIGC project (with HIPC resources)
Restructuring of ONADEF	Execution of ANAFOR working programme in accordance with the guidelines laid down by the reform of that institution
Combating illegal exploitation	Charging all established offences; application and collection of penalties Exclusion of all companies guilty of serious offences from submitting tenders for exploitation permits, and even suspension of valid permits in their keeping

2. Better environmental monitoring and implementation of international conventions and standards

The setting up of the two Ministries should allow for greater visibility and clarification of the roles of the different stakeholders. This should enable us tackle environmental problems unequivocally, given that the Permanent Secretariat for the Environment in the former MINEF was unable to go beyond the monitoring of a few conventions. The main mission here which consisted in setting the different standards in all sectors of activity was not carried out. Even the implementing instruments of the 1996 Framework law on the environment were not drafted. By setting up the Ministry of the Environment and Protection of Nature, the Government is thus endowing itself with an ambitious environmental policy that covers all sectors, and wards off the confusion that has always characterised environmental problems seen by the general public as limited to the Forest and Wildlife sector. The Ministry mainstreams all the sectors concerned so as to handle the cross-cutting nature of environmental actions.

3. The National Forest and Environment Sector Programme

This programme appears as a choice forum for the implementation of forestry policy during the next decade. Through the Forest and Environment Sector Programme, the Government seeks to secure multifaceted assistance for the effective implementation of its policy of sustainable and participatory management of forest resources, so that they contribute significantly to improve the national economy as well as the living environment and standards of the populations. Priority areas of such support include capacity-building of national institutions in terms of human, material and financial resources to enable them effectively implement the policy on the field. Such strengthening is to be achieved through implementation of institutional reforms envisaged within the framework of institutional review, and support of public and private efforts towards the sustainable management of forest and wildlife resources at the economic, ecological and social levels.

The Programme approach adopted for its preparation and implementation will also enable the Government to use FESP as a coordination framework for all actions that contribute to achieving the objectives of the country's forest and wildlife policy. FESP is thus conceived of by Government as a national sector development programme open to funding from all donors, the civil society, NGOs and the private sector. Through this programme, the Government seeks to endow itself with a control panel that will enable it pilot, monitor and control this sector with full knowledge of the situation on the ground. Funding of part of the programme in the form of budgetary support should also make it possible to build the technical and financial capacities of the administration.

The onus of implementation of FESP also lies with the three other groups of stakeholders that are the private sector, the civil society and donors. While the public sector will henceforth seek to concentrate on its traditional duties, the private sector will take care of production functions as a responsible partner concerned about upholding the laws and regulations in force. The civil society will participate in the implementation of FESP as partner and observer of compliance with the rules of the game by all stakeholders. Donors and development partners will participate in the implementation of FESP by harmonising their approaches and selecting the components they intend to finance and support, depending on their interests and programme cycles.

FESP is thus the framework-programme for the implementation of the forestry policy that essentially aims to ensure the conservation and sustainable management of forest ecosystems with a view to meeting the local, national, regional and world needs of present and future generations over the next ten (10) years, and aims to mobilise the possible contributions of various stakeholders including donors, NGOs and the civil society.

Funding of the Programme during its first years will be done on the basis of the results obtained and assessed with the triggers summarised in the table appended hereto.

Yaoundé, 1 March 2005

Minister for the Environment
and Protection of Nature
HELE Pierre

Minister for Forestry
and Wildlife
EGBE ACHUO Hillman

Impacts on Harvest Selectivity

FY species	1998–99		2002–03		Evolution	
	Vol. (m³)	Part (%)	Vol. (m³)	Part (%)	Variation (volume)	Variation (%)
Ayous/Obéché	718,165	37.06	706,432	36.27	–11,733	–1.63
Sappeli	469,019	24.20	408,521	20.98	–60,498	–12.90
Subtotal 2 species	*1,187,184*	*61.26*	*1,114,953*	*57.25*	*–72,231*	*–6.08*
Iroko	125,964	6.50	71,972	3.70	–53,992	–42.86
Fraké	89,775	4.63	84,341	4.33	–5,434	–6.05
Tali	81,971	4.23	136,190	6.99	54,219	66.14
Azobé	73,657	3.80	101,498	5.21	27,841	37.80
Subtotal 6 species	*1,558,551*	*80.43*	*1,508,954*	*77.48*	*22,634*	*–3.18*
Assamela	39,318	2.03	17,868	0.92	–21,450	–54.56
Sipo	26,867	1.39	34,951	1.79	8,084	30.09
Movingui	26,132	1.35	38,751	1.99	12,619	48.29
Dibétou	23,541	1.21	20,751	1.07	–2,790	–11.85
Ilomba	20,218	1.04	15,760	0.81	–4,458	–22.05
Moabi	18,326	0.95	28,630	1.47	10,304	56.23
Padouk blanc	18,263	0.94	3,796	0.19	–14,467	–79.21
Naga	17,644	0.91	2,991	0.15	–14,653	–83.05
Acajou de bassam	13,668	0.71	15,614	0.80	1,946	14.24
Doussié blanc	13,469	0.70	5,404	0.28	–8,065	–59.88
Kossipo	13,048	0.67	43,070	2.21	30,022	230.09
Lotofa/Nkanang	12,658	0.65	14,549	0.75	1,891	14.94

FY species	1998–99 Vol. (m³)	Part (%)	2002–03 Vol. (m³)	Part (%)	Evolution Variation (volume)	Variation (%)
Eyong	11,719	0.60	13,434	0.69	1,715	14.63
Ekaba	11,496	0.59	1,176	0.06	–10,320	–89.77
Subtotal 20 species	1,824,918	94.18	1,765,699	90.66	–9,622	–3.25
Doussié rouge	8,139	0.42	10,572	0.54	2,433	29.89
Aiélé/Abel	7,769	0.40	6,230	0.32	–1,539	–19.81
Faro	7,735	0.40	536	0.03	–7,199	–93.07
Bilinga	7,655	0.40	12,650	0.65	4,995	65.25
Fromager	7,212	0.37	15,281	0.78	8,069	111.88
Tali Yaoundé	6,606	0.34	2,452	0.13	–4,154	–62.88
Aningré R	5,038	0.26	3,758	0.19	–1,280	–25.41
Bété	4,939	0.25	4,793	0.25	–146	–2.96
Tiama	3,857	0.20	6,922	0.36	3,065	79.47
Subtotal 29 species	1,883,868	97.22	1,828,893	93.90	4,244	–2.92
Bubinga rose	3,847	0.20	1,160	0.06	–2,687	–69.85
Okan/Adoun	3,397	0.18	36,385	1.87	32,988	971.09
Koto	3,306	0.17	5,770	0.30	2,464	74.53
Onzabili/Angongui	3,004	0.16	4,933	0.25	1,929	64.21
Bossé foncé	2,787	0.14	2,173	0.11	–614	–22.03
Naga parallèle	2,497	0.13	627	0.03	–1,870	–74.89
Niové	2,324	0.12	1,230	0.06	–1,094	–47.07

FY species	1998–99		2002–03		Evolution	
	Vol. (m³)	Part (%)	Vol. (m³)	Part (%)	Variation (volume)	Variation (%)
Bubinga rouge	1,785	0.09	2,281	0.12	496	27.79
Mambodé/Amouk	1,712	0.09	2,956	0.15	1,244	72.66
Longhi/Abam	1,583	0.08	1,587	0.08	4	0.25
Gombé/Ekop ngombé	1,406	0.07	0	0.00	−1,406	−100.00
Dabéma	1,321	0.07	5,649	0.29	4,328	327.63
Kotibé	1,035	0.05	864	0.04	−171	−16.52
Mukulungu	1,015	0.05	2,313	0.12	1,298	127.88
Subtotal 43 species	1,914,887	98.82	1,896,821	97.39	36,909	−0.94
Production total	1,937,768	100.00	1,947,656	100.00	9,888	0.51

Source: Statistics on logging per species, SIGIF.

Evolution of Processing Capacity, 1998–2004

No.		Type of processing	Processing capacity (cubic meters)			Number of employees			Remarks
			1998	2004	Variation	1998	2004	Variation	
Mills in activity in 1998									
1	SEBC	SS	110,000	90,000	−20,000	128	145	17	
2	PROPALM	SS	50,000	50,000	0	102	0	−102	Mill closed
3	CIBC	SS	50,000	43,200	−6,800	130	95	−35	
4	J PRENANT	SS	38,500	54,000	15,500	111	130	19	
5	SAB	SS	35,000	35,000	0	98	0	−98	Mill closed
	Gr. THANRY		283,500	230,400	−53,100	569	370	−199	
6	SFID Mbang	S+S+M	150,000	150,000	0	587			
7	SFID Dimako	D	43,000	0	−43,000	325			Mill closed
8	SID	SS	60,000	60,000	0	105			
	Gr. ROUGIER		253,000	210,000	−43,000	1,017	950	−67	
9	SEFAC	S+S+M	130,000	96,000	−34,000	340	300	−40	
10	SEBAC	SS	36,000	50,000	14,000	47	150	103	
	Gr. SEFAC		166,000	146,000	−20,000	387	450	63	
11	ALPICAM	S+D+C	120,000	110,000	−10,000	696	888	192	
12	GRUMCAM	S+S	47,500	45,000	−2,500	150	424	274	
	Gr. ALPI		167,500	155,000	−12,500	846	1,312	466	
13	SFHSNT	S+S	25,000	0	−25,000	129	0	−129	Mill closed
14	SFHS	S+S	60,000	0	−60,000	230	0	−230	Mill closed
15	IBCAM	D	60,000	0	−60,000	384	0	−384	Mill ceased to Serrabocam

No.		Type of processing	Processing capacity (cubic meters)			Number of employees			Remarks
			1998	2004	Variation	1998	2004	Variation	
	Gr. HAZIM		145,000	0	-145,000	743	0	-743	
16	SFIL	SS	90,000	72,000	-18,000	148	300	152	
17	SOTREF	SS	ND	0	0		0	0	Mill ceased to FCA
18	MADEX	S+S	1,500	0	-1,500	50	0	-50	Mill transferred
	Gr. DECOLVENAERE		91,500	72,000	-19,500	198	300	102	
19	FOREST. DE CAMPO	SS	65,000	0	-65,000	177	0	-177	Mill ceased to SCIEB
20	SIBAF	S+S	71,000	0	-71,000	140	0	-140	Mill closed
	Gr. BOLLORE		136,000	0	-136,000	317	0	-317	
21	PALLISCO	S+S+M	42,000	55,000	13,000	126	135	9	
22	CIFM	SS	14,000	19,800	5,800	41	63	22	
	Gr. PASQUET		56,000	74,800	18,800	167	198	31	
23	EFMK	SS	30,000	30,000	0	98	98	0	
24	SABM	SS	13,000	30,000	17,000	40	40	0	Mill closed
25	SN COCAM	S+D+C	72,000	30,000	-42,000	170	40	-130	Not in activity (Sofopetra pb)
26	EGTF RC CORON	S+S	40,000	60,000	20,000	200	100	-100	
	Gr. KHOURY		155,000	150,000	-5,000	508	278	-230	Mills transferred to Khoury
27	TIB - SIM	S+séchoir+T	84,000	110,000	26,000	60	400	340	
28	PATRICE BOIS	S+séchoir	40,000	40,000	0		368	368	

No.		Type of processing	Processing capacity (cubic meters)			Number of employees			Remarks
			1998	2004	Variation	1998	2004	Variation	
29	CFE	SS	72,000	72,000	0	92	0	-92	Mill closed
30	IBC	S+séchoir	60,000	20,000	-40,000	500	130	-370	Scierie fermée - parqueterie fonctionne
31	SEFN	S+séchoir	48,000	48,000	0	140	0	-140	Mill closed
32	PK STF	SS	27,500	27,500	0	100	100	0	Mills transferred to Fadoul
33	TTS	SS	25,000	60,000	35,000	157	220	63	
34	SEEF	S+séchoir	22,000	35,000	13,000	140	173	33	
35	ECAM PLACAGES	T+S	42,000	16,500	-25,500	375	253	-122	
36	WIJMA	SS	70,000	105,000	35,000	217	370	153	
37	CFK	SS	15,000	15,000	0	97	0	-97	
38	DN KARAYANNIS	SS	18,000	18,000	0	65	65	0	
39	BTA	SS	50,000	0	-50,000	107	0	-107	Mill ceased to TRC
40	STIK	SS	25,000	0	-25,000	20	0	-20	Mill hired by TRC
41	SCTCB	SS	30,000	0	-30,000	54	0	-54	Mill closed
42	CAFOREX	SS	7,200	0	-7,200	30	0	-30	Mill closed
43	SOCAFI	SS	10,000	0	-10,000	50	0	-50	Mill closed
44	SIBOIS	SS	6,000	0	-6,000	30	0	-30	Mill closed
45	DESIGN	SS	10,000	0	-10,000	67	0	-67	Mill closed
	Subtotal		**2,115,200**	**1,647,000**	**-468,200**	**7,053**	**5,937**	**-1,116**	

No.	Type of processing	Processing capacity (cubic meters)			Number of employees			Remarks
		1998	2004	Variation	1998	2004	Variation	
New mills								
1	TRC Douala	SS+M	28,800	28,800		193	193	Rachat de BTA
2	TRC Kumba	SS+M	18,700	18,700		162	162	Mill hired to STIK
3	FILIERE BOIS	SS	50,000	50,000		150	150	Gr. SEFAC
4	CFC	S+S	43,200	43,200		130	130	Gr. THANRY
5	SCIEB	SS	30,000	30,000		96	96	Ceased by HFC/Bolloré
6	SCTB	S+D+C+T	84,000	84,000		357	357	
7	FIP CAM	SS	48,000	48,000		149	149	
8	INGF Ydé	S+S+M	110,000	110,000		180	180	
9	INGF Lomié	SS	90,000	90,000		180	180	
10	PLACAM	S+D+C	70,000	70,000		437	437	
11	SEFICAM	S+D	12,000	12,000		50	50	
12	IBCAM	D+C	60,000	60,000		150	150	Bought by Setrabocam
13	SEFICAM	S+D	12,000	60,000		70	70	
14	STBK	SS		ND				
15	MMG	SS		ND				
Subtotal			656,700	704,700	0	2,304	2,304	
TOTAL		2,115,200	2,303,700	236,500	7,053	8,241	1,188	

C: plywood, D: veneer, M: moulding, S: drying facility; S+S: sawmill with drying facility; SS: simple sawmill; T: Sliced-veneer

Plants open and closed during the period 1998–2004 : CALCO, EBIC, BOIS 2000, CANA BOIS, AGRO BOIS, FCA.

Source: Fochivé 2005

Modeling the Impact of Changing Fiscal Pressure

Common assumptions were made for three periods (1992, 1998, and 2004):

- Forest concession of 200,000 hectares (annually open area of 6,667 hectares)
- Average harvest intensity (cubic meters harvested per hectare) of 10 cubic meters/ha/year on licenses and on concessions.
- 30-year harvesting cycle
- Processing efficiency of 33 percent for exported sawnwood and 55 percent for veneer. In 1998 and 2005, the surtax on exported logs was assumed equal to CFAF 4,000 per cubic meter.

For 1992, taxes were calculated based on predefined values (*mercuriales*) for firms located in zone 3 (the most distant from ports). FOB prices for each period are from the international markets (B+ quality for logs, FAS for sawnwood, faces for veneers). The model has been established for the two major logged and exported species in Cameroon over the last decade, ayous and sapelli. Trends observed from the analysis of these two species are generalized to other species.

MODEL 1: Impacts on forest taxation by type of wood product

The tables below provide details of the calculation of the results presented in table 5.3.

Situation in 1992

Area fee	98	CFAF/ha
Felling tax	5%	of the mercuriales
Customs taxes	30%	of export mercuriales
Tax on log export	2%	of export mercuriales
ONADEF tax	8%	of export mercuriales

Ayous - Values			**Sapelli - Values**		
Mercuriales values 1992			*Mercuriales values 1992*		
Log	8,000	CFAF/m^3	Log	21,000	CFAF/m^3
Mercuriales export values 1992			*Mercuriales export values 1992*		
Log	11,000	CFAF/ m^3	Log	25,000	CFAF/m^3
FOB values 1992			*FOB values 1992*		
Log	47,500	CFAF/ m^3	Log	75,000	CFAF/m^3
			Sawnwood	140,000	CFAF/m^3
			Veneer	112,500	CFAF/m^3

Ayous - Taxes			**Sapelli - Taxes**		
Area fee			*Area fee*		
Total area fee	19,600,000	CFAF	Total area fee	19,600,000	CFAF
Area fee per volume	294	CFAF/m^3	Area fee per volume	294	CFAF/m^3
Felling tax			*Felling tax*		
Logs	400	CFAF/m^3	Logs	1,050	CFAF/m^3
Customs taxes			*Customs taxes*		
Logs	3,300	CFAF/m^3	Logs	7,500	CFAF/m^3
Tax on log export			*Tax on log export*		
Logs	220	CFAF/m^3	Logs	500	CFAF/m^3
ONADEF tax			*ONADEF tax*		
Logs	880	CFAF/m^3	Logs	2,000	CFAF/m^3
Total			**Total**		
Log	5,094	CFAF/m^3	Log	11,344	CFAF/m^3
	10.7%			15.1%	
			Sawnwood	4,073	CFAF/m^3 sawnwood
				2.9%	
			Veneer	2,444	CFAF/m^3 veneer
				2.2%	

Situation in 1998

Fiscal framework - 1998		
Area fee	1500	CFAF/ha
Felling tax	2.50%	of the log FOB value
Export tax	17.50%	of the log FOB value
	12.50%	of the log FOB value

Ayous - Values			**Sapelli - Values**		
FOB values 1998			*FOB values 1998*		
Log	80,750	CFAF/m^3	Log	121,125	CFAF/m^3
Sawnwood	203,000	CFAF/m^3	Sawnwood	292,000	CFAF/m^3
Veneer	270,000	CFAF/m^3	Veneer	270,000	CFAF/m^3
Ayous - Taxes			**Sapelli - Taxes**		
Area fee			*Area fee*		
Total area fee	300,000,000	CFAF	Total area fee	300,000,000	CFAF
Area fee per volume	4,500	CFAF/m^3	Area fee per volume	4,500	CFAF/m^3
Felling tax			*Felling tax*		
Logs	2,019	CFAF/m^3	Logs	3,028	CFAF/m^3
Export tax			*Export tax*		
Logs	14,131	CFAF/m^3	Logs	21,197	CFAF/m^3
Sawnwood	10,094	CFAF/m^3 sawnwood	Sawnwood	15,141	CFAF/m^3 sawnwood
Veneer	10,094	CFAF/m^3 veneer	Veneer	15,141	CFAF/m^3 veneer
Total			**Total**		
Log	20,650	CFAF/m^3	Log	28,725	CFAF/m^3
	25.6%			23.7%	
Sawnwood	29,848	CFAF/m^3 sawnwood	Sawnwood	37,953	CFAF/m^3 sawnwood
	14.7%			13.0%	
Veneer	21,946	CFAF/m^3 veneer	Veneer	28,828	CFAF/m^3 veneer
	8.1%			10.7%	

Situation in 2004

Fiscal framework - 2004		
Area fee	3000	CFAF/ha
Felling tax	2.50%	of the log FOB value
Sawmill entry tax	2.25%	of the log FOB value
Log export tax	17.50%	of the log FOB value

Ayous - Values			**Sapelli - Values**		
FOB values 2004			*FOB values 2004*		
Log	90,100	CFAF/m^3	Log	128,250	CFAF/m^3
Sawnwood	210,000	CFAF/m^3	Sawnwood	295,000	CFAF/m^3
Veneer	270,000	CFAF/m^3	Veneer	270,000	CFAF/m^3
Ayous - Taxes			**Sapelli - Taxes**		
Area fee			*Area fee*		
Total area fee	600,000,000	CFAF	Total area fee	600,000,000	CFAF
Area fee per volume	9,000	CFAF/m^3	Area fee per volume	9,000	CFAF/m^3
Felling tax			*Felling tax*		
Logs	2,253	CFAF/m^3	Logs	3,206	CFAF/m^3
Sawmill entry tax			*Sawmill entry tax*		
Logs	2,027	CFAF/m^3	Logs	2,886	CFAF/m^3
Tax on log export			*Tax on log export*		
Log export tax	15,768	CFAF/m^3	Log export tax	22,444	CFAF/m^3
Surtax	4,000	CFAF/m^3	Surtax	4,000	CFAF/m^3
Total			**Total**		
Log	33,048	CFAF/m^3	Log	41,536	CFAF/m^3
	36.7%			32.4%	
Sawnwood	40,242	CFAF/m^3 sawnwood	Sawnwood	45,733	CFAF/m^3 sawnwood
	19.2%			15.5%	
Veneer	24,145	CFAF/m^3 veneer	Veneer	18,349	CFAF/m^3 veneer
	8.9%			6.8%	

MODEL 2: Impacts of fiscal pressure on forest firms

Based on the results of Model 1, mean values were calculated for FOB prices and taxes, based on a 66 percent to 34 percent ratio of ayous to sapelli. The values used to calculate the results discussed in chapter 5 are provided in the table below.

	1992 mean*	1998 mean*	2004 mean*
FOB value - log	56,850	94,478	103,071
FOB value - sawnwood	140,000	233,260	238,900
FOB value - veneer	112,500	270,000	270,000
Taxation on log (m^3)	7,219	23,396	35,933
Taxation on sawnwood (m^3)	4,073	32,603	42,109
Taxation on veneer (m^3)	2,444	24,286	22,174
as percent FOB value for logs	12.2%	24.9%	35.2%
as percent FOB value for sawnwood	2.9%	14.1%	17.9%
as percent FOB value for veneer	2.2%	9.0%	8.2%

* Composition = ayous, 66%; sapelli, 44%

Technical Adaptations to Better Integrate the Tax Regime and Auction System

Experience over the past several years makes it possible to assess the pros and cons of the instruments used in Cameroon's forest tax regime and auction system and to suggest various adaptations that might improve their integration.

Area fee. The area fees proposed by private operators through the auctioning procedure are significantly higher than area fees fixed under the earlier administrative system. This change indicates that the industry's willingness to pay to access long-term concessions was higher than commonly thought. Competitively determined area fees

- Reveal the commercial timber value of forests and allow the country to capture a greater share of the forest economic rent;
- Allow more transparency in the allocation system and avoid disputes over the area fee level; and
- Allow local redistribution of the fees collected (to local councils and communities) on a simple and permanent basis (area based).

Relevance of preliminary inventory. One objection to the reform was that without a comprehensive inventory of resources throughout the forest, commercial concession value would remain largely unknown and the auction would be impracticable. Financial means should be given to the forest service (or to private firms acting on behalf of the forest service) to undertake *survey inventories* aiming at providing accurate *public information* of the commercial potential of the resource to be auctioned. In addition, sufficient time must be allowed for potential bidders to make their own surveys.

Floor price. Floor price matters, especially when competition is low or nonexistent. Experience has shown that when several bidders compete for a given concession, bids are well beyond the floor price. When no bids at all are received for a given concession, it is worth considering holding a new round with a lower floor price (a practice not authorized by current Cameroonian regulations).

Capture of economic rent. The auction system has been effective in capturing most of the economic forest rent, and both government and local council/community revenues (50 percent of the annual royalty) have increased. The structure of forest taxes has changed so that most are concentrated upstream; this trend is consistent with the decline of log exports following progressive implementation of the partial log export ban after 1999. Yet simply seeking to capture economic rent is not enough; this focus must be accompanied by tough policies to suppress the numerous dysfunctions of the administration that bring additional costs to the economic agents ("administrative taxes and fees," excessive requirements put within the "cahiers des charges," abusive financial cautions, excessive delays in refunding value added taxes for exporters, and so forth).

Export levies. Export levies are easiest to collect and offer fewer opportunities for fraud. They have a role to play even in regimes that emphasize upstream taxation. They can be used to promote more advanced processing capacity and the use of less-marketed species over less-processed goods made from more traditional species. The challenge is to calibrate export levies at levels that do not discourage exports (if too high), play their incentive role effectively, and allow adequate revenue capture by the state.

References

Abt, V., J.-C. Carret, R. Eba'a Atyi, and P. Mengin-Lecreulx. 2002. "Étude en vue de la définition d'une politique sectorielle de transformation et de valorisation du bois." CERNA-ONF International and ERE Développement, Centre d'économie Industrielle de l'Ecole des Mines de Paris, Paris.

Bawa, K. S., and R. Seidler. 1998. "Natural Forest Management and Conservation of Biodiversity in Tropical Forests." *Conservation Biology* 12: 46–55.

Biesbrouck, K. 2002. "New Perspectives on Forest Dynamics and the Myth of 'Communities.'" *IDS Bulletin* 33 (1): 55–64.

Bigombe Logo, P. 2008. *Foresterie communautaire et réduction de la pauvreté rurale au Cameroun: Bilan et tendances de la première décennie. World Rain-Forest Movement* no. 126 (January). http://www.wrm.org.uy/countries/Cameroon/Bigombe.html. Accessed 10 February 2009.

Binkley, C. S., and J. R. Vincent. 1992. "Forest-based Industrialization: A Dynamic Perspective." In N. Sharma, ed. *Managing the World's Forests.* Dubuque, Iowa: Kendall/Hunt.

Boscolo, M., and J. R. Vincent. 2007. "Impact of Area Taxes on Logging Behavior." *Environment and Development Economics* 12: 505–20.

Burnham, P. 2000. "Whose Forest? Whose Myth? Conceptualisation of Community Forest in Cameroon." In *Mythical Land, Legal Boundaries*, ed. A. Abramson and D. Theodossopoulos. London: Pluto Press.

Carret, Giraud, and Lazarus. 1998. "L'industrialisation de la filière bois au Cameroun entre 1994 et 1998. Observations, Interprétations, Conjectures." CERNA, Paris.

CBFP (Congo Basin Forest Partnership). 2006. *"Les forêts du Basin du Congo-Etat 2006."* http://www.cbfp.nyanet.de/cms/tl_files/archive/thematique/Les_forets_du_Bassin_du_Congo_etat_2006.pdf.

Cerutti, P. O., and L. Tacconi. 2008. "Forests, Illegality, and Livelihoods: The Case of Cameroon." *Society and Natural Resources* 21 (9): 845–53.

Chupezi, T. J., and O. Ndoye. 2006. "Commercialization of *Prunus africana* (African Cherry): Impacts on Poverty Alleviation in Cameroon." Presented at the workshop on Forestry, Wildlife, and Poverty Alleviation in Africa, March 27–28, Maputo, Mozambique.

CIRAD (Centre de Coopération Internationale en Recherche Agronomique pour le Développement) and I&D (Institutions et Développement). 2000. *Audit économique et financier du secteur forestier au Cameroun. Rapport final.* Yaoundé, Cameroon: Ministère de l'Economie et des Finances, Comité Technique de Suivi des Réformes Economiques.

———. 2006. *Audit économique et financier du secteur forestier au Cameroun.* Ministère de l'Economie et des Finances. Yaoundé, Cameroon.

Cuny, P., P. Abe'ele, N. Eboule Singa, A. Eyene Essomba, and R. Djeukam. 2004. *État des lieux de la foresterie communautaire au Cameroun.* U.K. Department for International Development (DFID), Yaoundé, Cameroon.

de Wasseige, C., D. Devers, P. de Marcken, R. Atyi Eba'a, R. Nasi, and Ph. Mayaux. 2009. *Les forêts du Bassin du Congo—état des Forêts 2008.* European Commission. Luxembourg: Publications Office of the European Union.

Economist magazine, February 14, 2008.

Fochivé, E. 2005. *Évolution du secteur forestier sur la période 1998–2003 (opérateurs économiques, investissements, emplois).* Background document for the 2006 *Audit économique et financier du secteur forestier au Cameroun.* Yaoundé, Cameroon.

Fredericksen, T. S., and F. E. Putz. 2003. "Silvicultural Intensification for Tropical Forest Conservation." *Biodiversity and Conservation* 12: 1445–53.

Geschiere, P. 2004. "Ecology, Belonging and Xenophobia: The 1994 Forest Law and the Issue of 'Community'." In *Rights and the Politics of Recognition in Africa*, eds. H. England and F. Nyamnjoh. New York: Palgrave Macmillan.

GFW (Global Forest Watch). 2000. *An Overview of Logging in Cameroon.* Washington, DC: World Resources Institute (WRI).

———. Various years. *Interactive Forestry Atlas of Cameroon* and *Overview.* www.globalforestwatch.org.

Godwin, O., and C. Tekwe. 1998. *Land Use Plan for Bimbia Bonadikombo.* Limbé, Cameroon: MCP.

Greenpeace. 2007. "Forest Reforms in the DRC: How the World Bank is Failing to Learn the Lessons from Cameroon." Amsterdam: Greenpeace International. http://www.greenpeace.org/usa/assets/binaries/lessons-learned-from-cameroon.

GTZ (Deutsche Gesellschaft für Technische Zusammenarbeit). 2006. *Étude comparative de 20 plans d'aménagement approuvés au Cameroun.* Programme de Gestion Durable des Ressources Naturelles (PDGRN), GTZ Cameroon. November.

GTZ (Deutsche Gesellschaft für Technische Zusammenarbeit) and FAO (Food and Agriculture Organization). 2007. *Commerce sous-régional et international des produits forestiers non ligneux alimentaires et des produits agricoles traditionnels en Afrique Centrale.* Eschborn and Rome.

Handja, G. T. 2007. *La reconnaissance des droits des communautés Pygmées du Sud Cameroun sur les resources naturelles.* Presented at the Rights and Resources Initiative workshop, Douala, Cameroon, December 2007.

I&D (Institutions et Développement). 2002. "Rapport de la revue institutionnelle du secteur forestier au Cameroun." Yaoundé, Cameroon.

Jackson, D. 2004. "Implementation of International Commitments on Traditional Forest-related Knowledge: Indigenous Peoples' Experiences in Central Africa." Prepared for Expert Meeting on Traditional Forest-Related Knowledge, December 6–10, San Jose, Costa Rica.

Julve, C., M. Vandenhaute, C. Vermeulen, B. Castadot, H. Ekodeck, and W. Delvingt. 2007. "Séduisante théorie, douloureuse pratique: la foresterie communautaire camerounaise en butte à sa propre legislation." *Parcs et Réserves* 62 (2): 18–24.

Karsenty, A., 2007. Questioning rent for development swaps: new market-based instruments for biodiversity acquisition and the land-use issue in tropical countries. *International Forestry Review* 9 (1): 503–513

Karsenty, A., and S. Gourlet-Fleury. 2006. "Assessing Sustainability of Logging Practices in the Congo Basin's Managed Forests: The Issue of Commercial Species Recovery." *Ecology and Society* 11(1): 26. http://www.ecologyandsociety.org/vol11/iss1/art26/.

Karsenty, A., L. Mendouga, and A. Pénelon. 1997. "Spécialisation des espaces ou gestion intégrée de massifs? Le cas de l'Est-Cameroun." *Bois et Forêts des Tropiques* 251: 43–53.

Kim, Noëlle Brice. 2007. "Avant nous avions les yeux clos, maintenant nos yeux sont ouverts. Maintenant je m'exprime. Ce n'était pas le cas avant." Presented at the Rights and Resources Initiative workshop, Douala, Cameroon, December 2007.

Koffi Yeboah, Alexis. 2005. "Sciage artisanal, transformation et commerce du bois d'œuvre du Cameroun à destination de l'arc soudano-sahélien." CNEARC-ENGREF-CIRAD.

Kuetche, M. 2006. *Contribution à l'audit économique du secteur forestier.* National Statistical Institute of Cameroon, Yaoundé, Cameroon.

Lescuyer, G., H. Ngoumou Mbarga, and P. Bigombe Logo. 2008. "Use and Misuse of Forest Income by Rural Communities in Cameroon. " *Forests, Trees, and Livelihoods* 18: 291–304.

Malleson, R. 2001. *Perspectives et contraintes en matière de gestion communautaire des forêts: conclusions d'études effectuées dans la forêt de Korup, dans la province du sud-ouest, au Cameroun.* Published in Document Réseau de Foresterie pour le Développement Durable No. 25, http://www.odi.org .uk/networks/rdfn/rdfn-25a-francais.pdf.

Mallet, B., F. Besse, D. Gautier, D. Muller, N. Bouba, and C.F. Njiti. 2003. "Quelles perspectives pour les gommiers en zone de savanes d'Afrique centrale?" In *Savanes africaines: des espaces en mutation, des acteurs face à de nouveaux défis,* ed. J.-Y. Jamin, L. Seiny Boukar, and C. Floret. Minutes of a colloquium, May 27–31, 2002, Garoua, Cameroon. Montpellier: Centre de Coopération Internationale en Recherche Agronomique pour le Développement (CIRAD). CD-Rom.

Mayaux, P., P. Defourny, D. Devers, M. Hansen, and G. Duveiller. 2007. "The Forests of the Congo Basin: State of the Forests, 2006." Document prepared for the Congo Basin Forest Partnership available online at http://www.cbfp .org/congobasin.html

Mefoude, S. 2007. "Des Bagyéli propriétaires terriens." *Bubinga* 117: 6–7.

Milgrom, P. R., and R. J. Weber. 1982. "A Theory of Auctions and Competitive Bidding." Part I. *Econometrica* 50: 1089–1122.

Nelson, J. 2007. "Securing Indigenous Land Rights in the Cameroon Oil Pipeline Zone." Forest People Programme, London.

Nguenang, G.M., Q. Delvienne, V. Beligne, and M. Mbolo. 2007. *La gestion décentralisée des ressources forestières au Cameroun: Les forêts communales après les forêts communautaires.* Presented at the 6th Conference on Central African Moist Forest Ecosystems (CEFDHAC), November 20–23, Libreville, Gabon.

Oyono, R. 2005. "Profiling Local-Level Outcomes of Environmental Decentralizations: The Case of Cameroon's Forests in the Congo Basin." *Journal of Environment and Development* 14: 317–37.

Plouvier, D., R. Eba'a Atyi, T. Fouda, R. Oyono, and R. Djeukam. 2002. *Etude du sous-secteur sciage artisanal au Cameroun.* AGRECO, Department for International Development, and PSFE, Yaoundé, Cameroon.

Poore, D., P. Burgess, J. Palmer, S. Rietbergen, and T. Synnott. 1989. *No Timber without Trees.* London: Earthscan.

REM (Resource Extraction Monitoring). 2007. "Independent Monitoring: Progress in Tackling Illegal Logging in Cameroon, 2006–2007." Annual Report. Cambridge, U.K: REM.

Takforyan, A. 2001. *"Chasse villageoise et gestion locale de la faune sauvage en Afrique: Une étude de cas dans une forêt de l'Est-Cameroun."* PhD thesis. Ecole des Hautes Etudes en Sciences Sociales, Paris.

Tchuitcham, G. 2006. *"Décentralisation et gestion communautaire des forêts au Cameroun: Opportunités et contraintes pour le développement local et la gestion durable des resources."* Thesis. Département des Sciences et Gestion de l'Environnement, Université de Liège.

Transparency International. 2007. "Cameroon 2007." National Integrity Systems Country Study Report. Berlin. http://www.transparency.org/policy_research/nis/regional/africa_middle_east.

———. Various years. Corruption Perception Index. Berlin. http://www.transparency.org/policy_research/surveys_indices/cpi.

UNDP (United Nations Development Programme). 2003. *Republic of Cameroon MDG's Progress Report at a Provincial Level.* Yaoundé, Cameroon: UNDP.

Vermeulen, C., and S. M. Carrière. 2001. *"Stratégies de gestion des ressources naturelles fondées sur les maîtrises foncières coutumières."* In *L'homme et la forêt dense humide tropicale,* ed. W. Delvingt, pp. 109–44. Gembloux, Belgium: Presses Agronomiques de Gembloux, Faculté des Sciences agronomiques de Gembloux.

Vincent, J. R., C. C. Gibson, and M. Boscolo. 2005. "The Politics and Economics of Timber Reforms in Cameroon." Washington, DC: World Bank Institute/World Bank.

WRI (World Resources Institute). 2000. *The Right Conditions: The World Bank, Structural Adjustment, and Forest Policy Reform.* Washington, DC: WRI.

Yale Center for Environmental Law and Policy and CIESIN (Center for International Earth Science Information Network, Columbia University). 2005. *2005 Environmental Sustainability Index: Benchmarking National Environmental Stewardship.* New Haven, CT: Yale. http://www.yale.edu/esi/ESI2005_Main_Report.pdf.

Index

ECO-AUDIT
Environmental Benefits Statement

The World Bank is committed to preserving endangered forests and natural resources. This book, *The Rainforests of Cameroon: Experience and Evidence from a Decade of Reform,* is printed on 60# Rolland Opaque, a recycled paper made with 30 percent post-consumer waste. The Office of the Publisher follows the recommended standards for paper usage set by the Green Press Initiative, a nonprofit program supporting publishers in using fiber that is not sourced from endangered forests. For more information, visit www.greenpressinitiative.org.

Savings:

- 7 trees
- 4 million BTUs of total energy
- 492 pounds of net greenhouse gases
- 2,202 gallons of waste water
- 260 pounds of solid waste

Map 1. Vegetation Cover in Cameroon

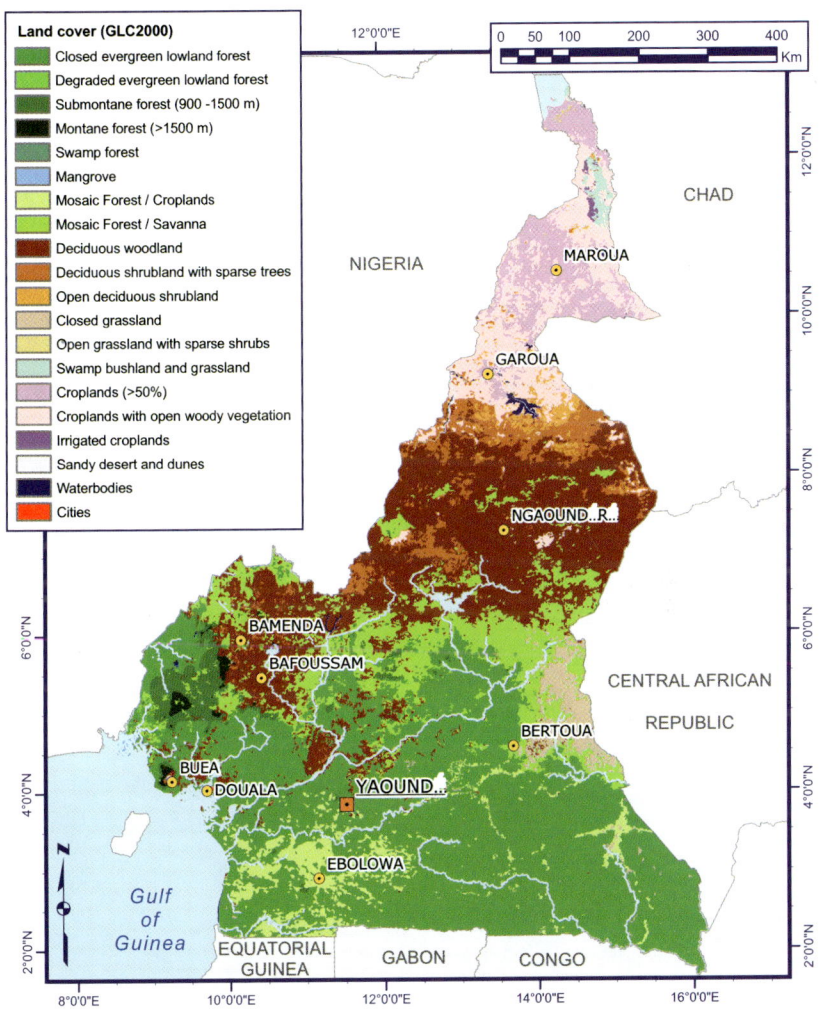

Land cover (GLC2000)
- Closed evergreen lowland forest
- Degraded evergreen lowland forest
- Submontane forest (900 -1500 m)
- Montane forest (>1500 m)
- Swamp forest
- Mangrove
- Mosaic Forest / Croplands
- Mosaic Forest / Savanna
- Deciduous woodland
- Deciduous shrubland with sparse trees
- Open deciduous shrubland
- Closed grassland
- Open grassland with sparse shrubs
- Swamp bushland and grassland
- Croplands (>50%)
- Croplands with open woody vegetation
- Irrigated croplands
- Sandy desert and dunes
- Waterbodies
- Cities

Sources: Vegetation cover: GLC2000 (EU Joint Research Centre, 2003); Cameroon administrative boundaries, hydrology, settlements, and coastline: Interactive Forestry Atlas of Cameroon, Version 2 (Cameroon Ministry of Forestry and Wildlife / World Resources Institute, 2007).

Map 2. Forest Use in Cameroon, 1992–2007

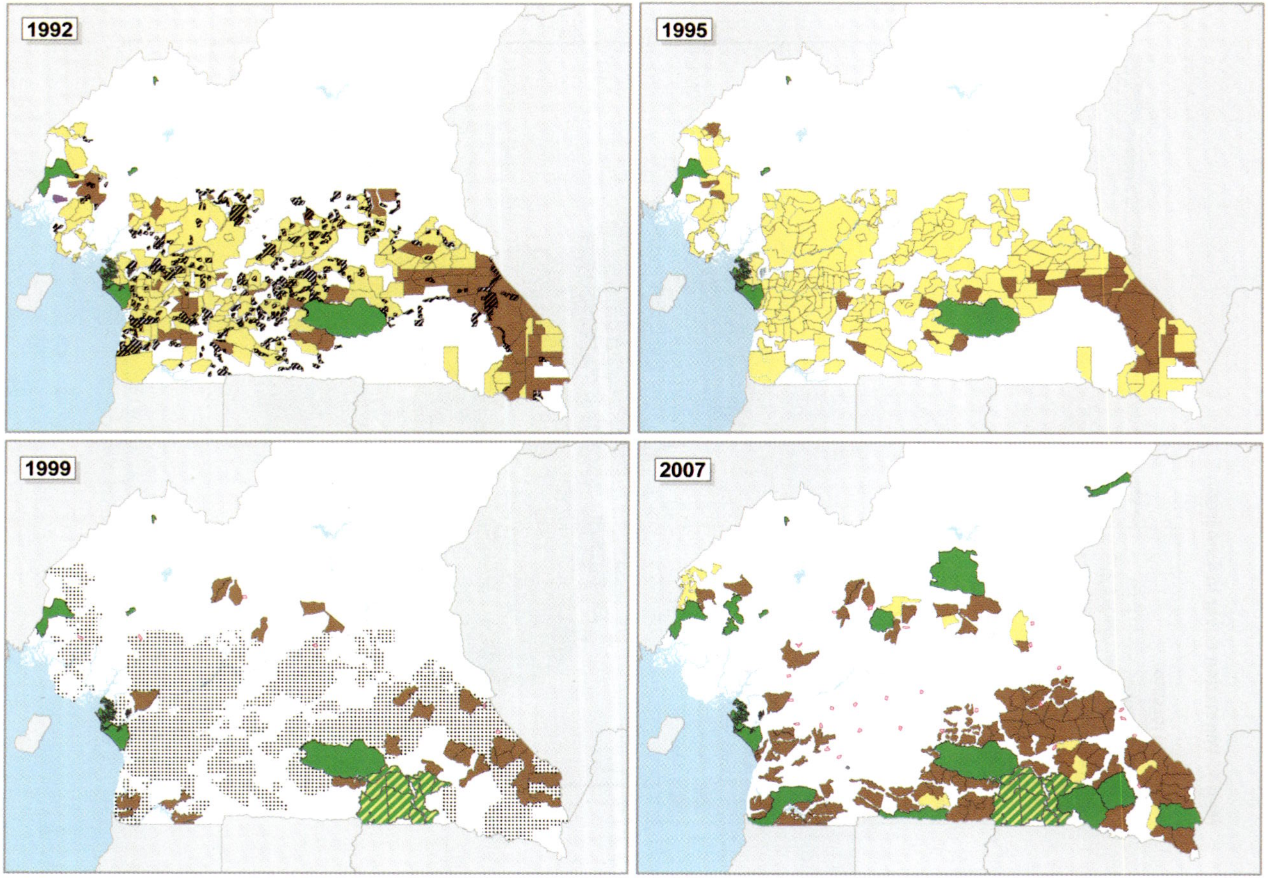

Valid logging titles

Licenses: 1992, 1995 FMUs[1]: 1999, 2007

	1992		1995		1999		2007	
	Area (hectares)	% of total area	Area (hectares)	% of total area	Area (hectares)	% of total area	Area (hectares)	% of total area
Active	2,898,700	22.1	1,910,400	16.8	1,693,600	99.1	5,557,200	88.6
Not active	7,155,000	64.5	9,458,000	83.2	—	—	642,500	10.2
Status unknown	20,100	0.2	—	—	—	—	—	—
Petits titres[2]	1,723,600	13.2	Unknown[3]	Unknown[3]	—	—	Unknown[3]	Unknown[3]
Sales of standing volume (SSV)	—	—	—	—	15,900[4]	0.9[3]	71,200	1.1
	11,797,500	100%	**11,368,400**	100%	**1,709,500**	100%	**6,270,900**	100%
Conservation concessions	—	—	—	—	—	NA	—	NA
Protected areas (IUCN categories I–IV)					867,000		867,000	
Abandoned licenses								

Data sources: Logging titles and other uses (1995, 1999, and 2007): Interactive Forestry Atlas of Cameroon, Version 2 (Cameroon Ministry of Forestry and Wildlife / World Resources Institute, 2007). Forest licenses and *petits titres* (1992): UNEP_WCMC, 1996, digitized by WRI, 2008. Protected areas: Interactive Forestry Atlas of Cameroon, Version 2 (Cameroon Ministry of Forestry and Wildlife / World Resources Institute, 2007), updated with "fiche synthétique des aires protégées du Cameroun" ["Summary of protected areas in Cameroon"] (FORAC, 2007) and World Database on Protected A eas (UNEP-WCMC, 2007). Cameroon administrative boundaries, hydrology, settlements, and coastline: Interactive Forestry Atlas of Cameroon, Version 2 (Cameroon Ministry of Forestry and Wildlife / World Resources Institute, 2007).

Notes:
[1] Forest Management Units.
[2] *petits titres* overlap with licences.
[3] *petits titres* not available.
[4] Data incomplete.
— Unavailable
NA Not applicable

Map 3. Status of Logging Titles in 2007

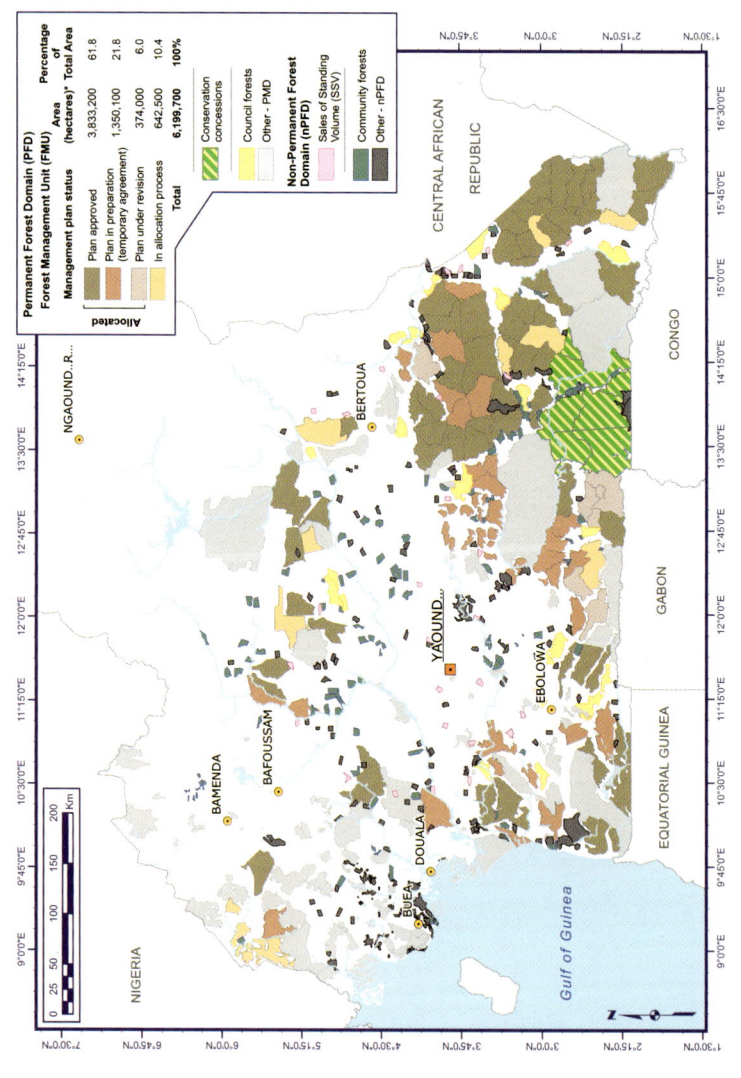

Permanent Forest Domain (PFD)		
Forest Management Unit (FMU)		
Management plan status	Area (hectares)*	Percentage of Total Area
Plan approved	3,833,200	61.8
Plan in preparation (temporary agreement)	1,350,100	21.8
Plan under revision	374,000	6.0
In allocation process	642,500	10.4
Total	**6,199,700**	**100%**

Conservation concessions

Council forests

Other – PMD

Non-Permanent Forest Domain (nPFD)

Sales of Standing Volume (SSV)

Community forests

Other – nPFD

Data sources: FMUs, SSVs, community forests, council forests and other uses: Interactive Forestry Atlas of Cameroon, Version 2 (Cameroon Ministry of Forestry and Wildlife/World Resources Institute, 2007); Cameroon administrative boundaries, hydrology, settlements, coastline: Interactive Forestry Atlas of Cameroon, Version 2 (Cameroon Ministry of Forestry and and Wildlife/World Resources Institute, 2007).

Map 4. Timber Export Routes and Volumes

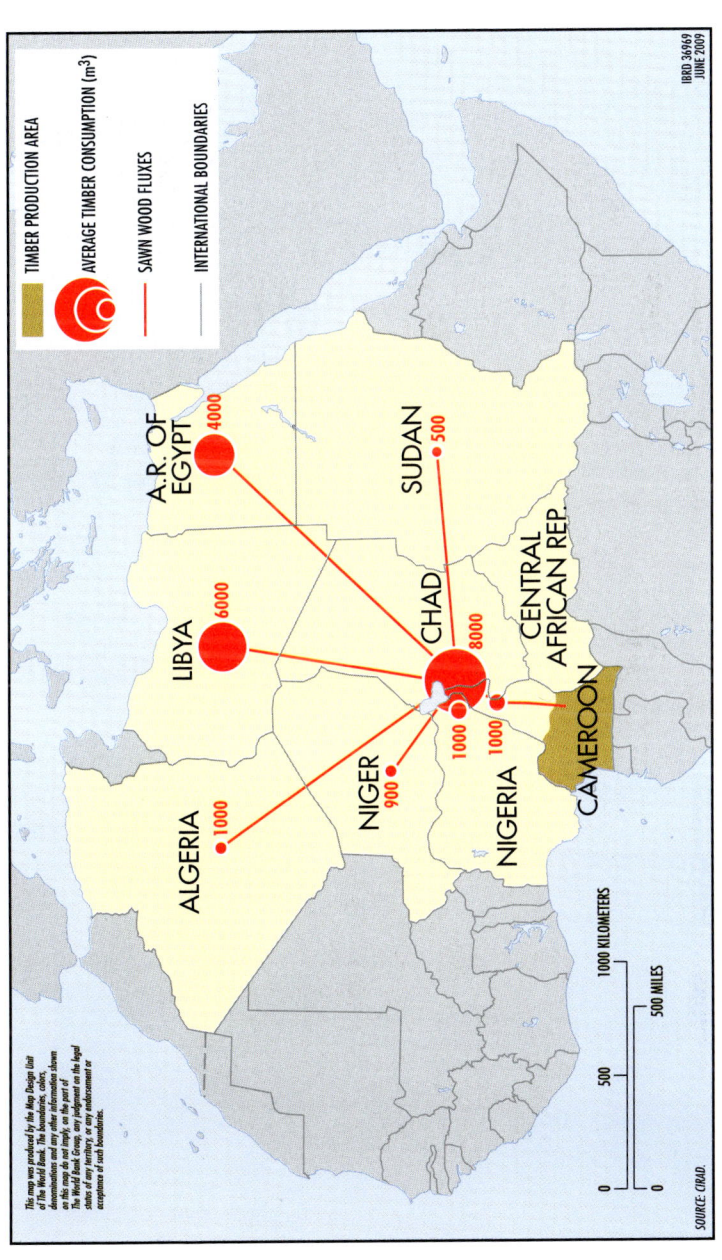

LEGEND

▮ TIMBER PRODUCTION AREA

◉ AVERAGE TIMBER CONSUMPTION (m³)

— SAWN WOOD FLUXES

— INTERNATIONAL BOUNDARIES

A.R. OF EGYPT · 4000

LIBYA · 6000

ALGERIA · 1000

SUDAN · 500

CHAD · 8000

NIGER · 900

NIGERIA · 1000 · 1000

CENTRAL AFRICAN REP.

CAMEROON

0   500   1000 KILOMETERS

0   500 MILES

SOURCE: CIRAD.

IBRD 36969
JUNE 2009

This map was produced by the Map Design Unit of The World Bank. The boundaries, colors, denominations and any other information shown on this map do not imply, on the part of The World Bank Group, any judgment on the legal status of any territory, or any endorsement or acceptance of such boundaries.

Source: CIRAD.

CAR = Central African Republic

Map 5. Status of Protected Areas in 1995 and 2008

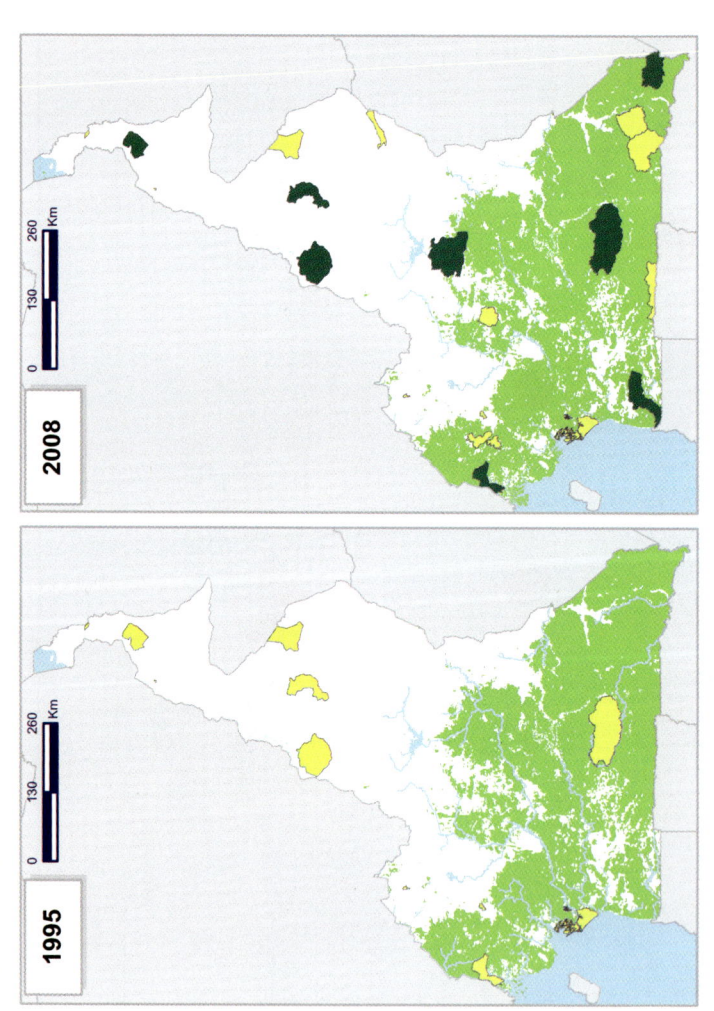

Protected Areas (IUCN categories I–IV)

Management plan adoption	1995		2008	
	Area (hectares)	% of total area	Area (hectares)	% of total area
■ Adopted	0	0.0	2,321,100	62.4
■ Not adopted yet	1,820,700	100.0	1,397,700	37.6
Total	1,820,700	100%	3,718,800	100%

■ Extent of forested areas

☐ Extent of nonforested areas

Data sources: Protected areas and management plan information: Interactive Forestry Atlas of Cameroon, Version 2 (Cameroon Ministry of Forestry and Wildlife / World Resources Institute, 2007), updated in 2008 by WRI, with "Fiche synthétique des aires protégées du Cameroun" ["Summary of protected areas in Cameroon"] (FORAC, 2007), the World Database on Protected Areas (UNEP-WCMC, 2007), and communications with the Cameroon Ministry of Forestry and Wildlife in May 2008; Forest extent: GLC2000 (EU Joint Research Centre, 2003); Cameroon administrative boundaries, hydrology, and coastline: Interactive Forestry Atlas of Cameroon, Version 2 (Cameroon Ministry of Forestry and Wildlife / World Resources Institute, 2007).

Map 6. Conservation of the Ngoila-Mintom Rainforests

NGOYLA-MINTOM CONSERVATION CONCESSION COMPLEX

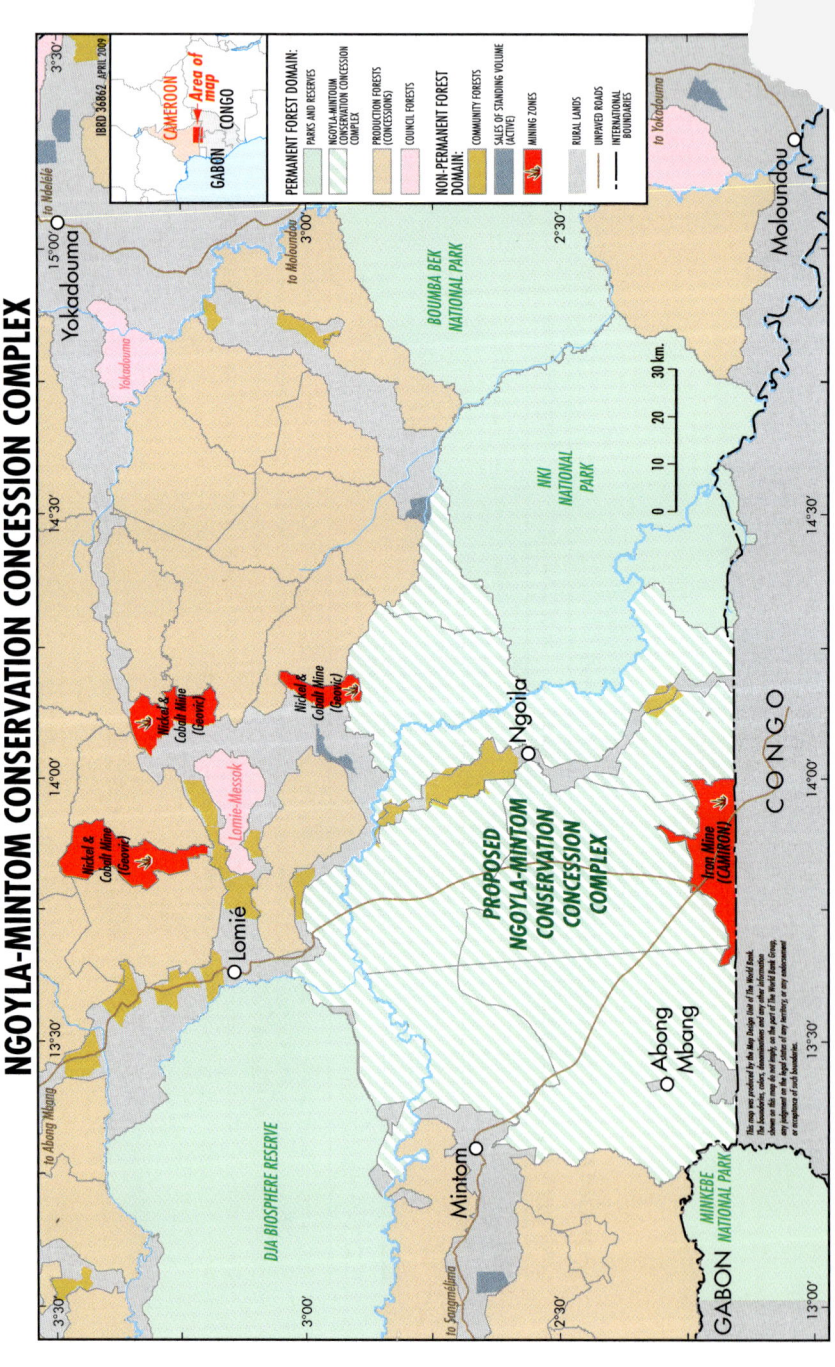

IBRD 36862 APRIL 2009

CAMEROON

Area of map

GABON   CONGO

PERMANENT FOREST DOMAIN:

- PARKS AND RESERVES
- NGOYLA-MINTOM CONSERVATION CONCESSION COMPLEX
- PRODUCTION FORESTS (CONCESSIONS)
- COUNCIL FORESTS

NON-PERMANENT FOREST DOMAIN:

- COMMUNITY FORESTS
- SALES OF STANDING VOLUME (ACTIVE)
- MINING ZONES
- RURAL LANDS
- UNPAVED ROADS
- INTERNATIONAL BOUNDARIES

to Ndélélé

to Yokadouma

Yokadouma

to Moloundou

3°30'

3°00'

2°30'

Moloundou

Yokadouma

BOUMBA-BEK NATIONAL PARK

to Moloundou

NKI NATIONAL PARK

0   10   20   30 km.

Nickel & Cobalt Mine (Geovic)

Nickel & Cobalt Mine (Geovic)

Ngoila

Nickel & Cobalt Mine (Geovic)

Lomié-Messok

Lomié

PROPOSED NGOYLA-MINTOM CONSERVATION CONCESSION COMPLEX

Iron Mine (CAMIRON)

CONGO

to Abong Mbang

Abong Mbang

DJA BIOSPHERE RESERVE

Mintom

to Sangmélima

GABON

MINKEBE NATIONAL PARK

to Yokadouma

3°30'   3°00'   2°30'

15°00'   14°30'   14°00'   13°30'   13°00'

This map was produced by the Map Design Unit of The World Bank. The boundaries, colors, denominations and any other information shown on this map do not imply, on the part of The World Bank Group, any judgment on the legal status of any territory, or any endorsement or acceptance of such boundaries.